Museum Collections and Today's Computers

MUSEUM COLLECTIONS AND TODAY'S COMPUTERS

Robert G. Chenhall *and*
David Vance

GP

GREENWOOD PRESS

NEW YORK
WESTPORT, CONNECTICUT
LONDON

Library of Congress Cataloging-in-Publication Data

Chenhall, Robert G., 1923–
Museum collections and today's computers /
Robert G. Chenhall and David Vance.
p. cm.
Bibliography: p.
Includes index.
ISBN 0-313-25339-0 (lib. bdg. : alk. paper)
1. Museums—Data processing. I. Vance, David, 1926–
II. Title.
AM139.C485 1988
069.5′0285—dc19 88-3091

British Library Cataloguing in Publication Data is available.

Library of Congress Catalog Card Number: 88-3091
ISBN: 0-313-25339-0

First published in 1988

Greenwood Press, Inc.
88 Post Road West, Westport, Connecticut 06881

Printed in the United States of America

The paper used in this book complies with the Permanent Paper Standard issued by the National Information Standards Organization (Z39.48-1984).

10 9 8 7 6 5 4 3 2 1

Contents

Figures

Preface

Thirteen years ago, the present authors were both involved in the preparation of a book called *Museum Cataloging in the Computer Age* (see Chenhall 1975). Since that book was published, computers have changed so drastically that it is like comparing the Model-T Ford to the luxury sports sedans of today. Almost all computer input in 1975 was in the form of punched cards that were delivered in "batches" for processing overnight on the large machines that only universities and other major institutions could afford; and the universal form of output was the printed pages that could be picked up the following morning— if there had not been a breakdown during the night (something that was very common). Who could have guessed that within such a few years we could all have available on our desks as much or more computing power than even those large machines possessed then.

Because of the dramatic changes in the computers available, it was not possible to consider re-writing and updating the earlier book. The present volume includes approaches, concepts, and techniques that are completely different from our ways of thinking 13 years ago. The principles of cataloging have not changed significantly since 1975, but the machines available to assist in the creation of usable museum catalogs are nothing short of revolutionary.

It is our intention with this book to serve three different audiences. First, we want to give the executives of major museums a better understanding of the kinds of detailed information museum employees need in order to function efficiently and the computer facilities that are necessary to provide that information.

Our second audience is the large number of persons—whatever their titles:

curator, director, volunteer—who have the sole responsibility, or nearly so, for managing and controlling our numerous small museums. We hope these people will discover what can be accomplished with the relatively inexpensive and easy-to-use microcomputers that are on the market today.

Finally, we trust that computer scientists, programmers, systems analysts, librarians, and others who may have a peripheral interest in the use of computers in museums will be able to learn about the particular and unusual information demands that stem from the control of museum objects in order to serve their museum clients well.

In Part 1, we focus on the unusual and not well-understood nature of the artifact records a museum should maintain, noting particularly the similarities and differences relative to library catalog records. Then, after a brief discussion of computer basics, in Part 2 we go into the details of data base management systems (DBMSs)—the only kind of computer programs capable of processing the files of data that every museum, large or small, must have in order to provide its employees with the collections information they need to do their jobs. The discussion is about principles, and no recommendation is made that any DBMS is patently superior to any other. However, in Chapter 13 there is a brief consideration of the problems of selection together with a list of some of the systems that have been used successfully by museums (see figure 15).

Other related problems are considered in Parts 3 and 4, and in the last chapter, the reader is taken through the step-by-step design and implementation of a computerized catalog file for a small museum, using a microcomputer and a DBMS designed for that computer. An extensive glossary of terms and a list of references is included at the end of the book.

PART I

MUSEUMS, ARTIFACT RECORDS, AND COMPUTERS

1

The World of (Almost) Unique Objects

Electronic computers can be useful tools in the performance of many different functions within a nonprofit institution, just as they are now considered the basic implements of word processing, accounting, graphics design, and other activities in business, scientific, and other types of organizations. The application that is distinct to museums and that will be emphasized throughout this book, however, is the use of computers to record and control activities having to do with physical objects that comprise museum collections: accessioning, registration, cataloging, inventorying, lending, exhibiting, research, etc. The use of now relatively inexpensive computers to store, process, and retrieve the verbal (and sometimes visual) symbols that represent artifacts in a museum's collection can greatly aid in the performance of these essential activities. In fact, even in modest-sized museums, it is now virtually impossible (a good case can be made for it having always been so) to maintain control of museum collections and activities without tools such as the computer.

Having made a case for the use of computers, we must quickly add the admonition that the computer alone will not automatically bring the collections of any museum under control. In 20 years of work in this field, we have both seen numerous cases of hopes dashed and money misspent pursuing the elusive dream of a "perfect" system that will take artifact records created by many different people over 30 years or more and distill them into exactly what is needed to run a museum. We must constantly remind ourselves that (1) it is only human intelligence that makes a computer function at all, (2) it is only museum people who can make a computer perform so as to accomplish the needs of museums, and (3) it is only the individuals in particular museums who make it possible to utilize a computer to accomplish the particular information needs of that institution.

Most of the mistakes that have been made over the years have been the result of smart, sometimes brilliant, computer scientists designing systems to accomplish what *they* thought museums needed in the way of information to perform their functions. Often this occurred not as a result of arrogance on the part of the computer scientists—though this has at times interfered as well—but rather because the museum people with whom they were working were either unable to think in the precise, consistent terms demanded by computer technology or were unwilling to discipline themselves sufficiently to think through the kinds of information they really needed to do their jobs. "Give us everything you can, we might need it some day," is an all too typical response illustrating what we will sometimes do to avoid really hard thinking.

Somewhat related to the desire of the computer scientist to create the perfect museum system is the frequent suggestion that museums should adopt a computer system that has demonstrated its utility in another discipline, perhaps medicine, law, or, most often, the library field. To the uninitiated, it seems entirely logical that since libraries have highly developed schemes for the recording of physical objects—books—it should be simple to adapt one of these schemes for the recording of museum collections. The structure of a library catalog record is theoretically quite similar to the structure of a museum catalog record, so an understanding of the former provides a good starting point for understanding the latter.

Years ago, through the activities of committees composed of members of the ALA (American Library Association) and similar organizations, librarians agreed among themselves that most of the time people search for books in only one of three ways: by author name, title, or subject. The complete catalog record on any book, of course, requires additional information such as publisher and date and place of publication, but these committees reasoned correctly that if they recorded a precise but limited quantity of data on 3″ x 5″ cards and filed one copy of the card in alphabetical order by author, another copy in alphabetic order by title, and a third according to subject, they would then have resolved for all time the needs of library patrons in searching for library books. In later years, refinements of this basic system have been made and two subject classification schemes are now in use, the Dewey system and the Library of Congress (LC) system. However, the initial scheme has worked and continues almost unchanged to this day.

It has been recognized for a long time that the most difficult and expensive part of library cataloging was the cost of having a specialist prepare the initial record to begin with. Reasoning that this would not have to be repeated over and over in all the libraries in the country if a means could be found to do it once and reproduce catalog cards to send to libraries led to the Library of Congress becoming the central source of cataloging data. Today, all major publishers provide essential catalog data shown on the inside cover of every book as "Library of Congress Cataloging in Publication Data."

Whether the catalog in a particular library is made available in the form of 3″ x 5″ cards filed manually by author, title, and subject or through some form of electronic search device, the basic record on each book consists of a limited number of *fields* of data (terms that may or may not be familiar to the reader are included in an extensive glossary at the end of the book). For each such field, there is a precisely defined method for recording information: alphabetically, numerically (perhaps with a limited number of digits, and possibly in a tightly structured data code form), in a prescribed syntactic arrangement, or using only words or phrases that are listed in an authority file such as a word list or thesaurus. In other words, even though the cataloging of a book may require only a few fields of data, both the fields and the content of what is recorded are highly structured. A moment of reflection regarding how these files are used reveals why this tight structuring and control of the input was desirable even in the days when the records were prepared manually (including on typewriters); with computers, however, such structure and control is essential because there is no longer a human element intervening to pick up misspellings or errors in content. Catalog files contain records specifically intended to be placed in some kind of order, usually alphabetic according to the contents contained in one or more fields. A computer is essentially a "dumb" machine that with mechanical dependability looks at each letter, number, character, or blank space in sequence and performs functions such as searching, sorting, or filing on the basis of what is found (more on this in Chapter 3). Therefore, if the ultimate retrieval of needed records is to be accurate and complete (and retrieval is what this kind of file is all about), it is essential that the contents of each record be structured the same way and have dependable terminology, word order, spelling, and meaning.

If libraries have been so successful in converting their catalog records from typed cards to electronic storage, why have museums been so slow to adopt more modern technology? There are several answers to this question, each of them correct in itself but accounting for only part of the problem.

The basic reason for maintaining library catalog files and museum artifact catalog files is the same: to assist in locating physical objects by providing written records that represent those objects. The physical objects in libraries, though, are sufficiently uniform that it was possible long ago to develop standardized systems for the storage of the objects and for their representation in written records. Library catalogs, basically, are quite simple: each record contains a limited number of precisely defined fields of information, and searching the file of records is usually done according to no more than three of these fields (author, title, and subject). To some extent, this simple structure is arbitrary, created by the common consent of librarians, but for the most part it serves the needs of library patrons adequately.

By contrast, those who use museum collections (not the visiting public, but museum registrars, conservators, curators, and scientists) differ among themselves in what they consider to be the important attributes of an artifact or specimen. The registrar, for example, is primarily interested in inventory control, which involves fields of information such as storage location, object number and name, source (donor, vendor, etc.), condition, and valuation. The scientist, on the other hand, will normally concentrate on an in-depth study of only a small part of the collection, which, in an archaeological collection for example, requires the recording of attributes such as the precise location where an artifact was originally found, by whom and with what associations, the material(s) out of which it is made, and implied techniques of fabrication. Because these different groups of users see different purposes for maintaining artifact files, the usual practice is to attempt to include *all* of the different fields of data that *may* be important to *any* of the probable users in the initial record.

If one adds to this complexity the wide diversity in types of museums and the differences between what is important, for example, in a university geological museum as contrasted with a large, public, fine arts museum, it is easy to see why progress in standardization has been so slow. The inevitable result has been an almost complete lack of uniformity among museums in what should be included in a catalog record, other than in situations such as within the National Park Service, where a governmental authority can dictate what information a number of museums must maintain for inventory control purposes, or where there are special files developed because of comparable interest within a particular part of the scientific community. In sum, the multiple purposes served by museum catalog files lead inevitably to the recording of many artifact attributes in place of the no more than six to ten attributes sufficient to catalog a book, and to a complexity of catalog records unknown to the library profession.

As a further extension of these differences, we will see when we come to the discussion of electronic networking in Chapter 12 that library science is a discipline ideally suited to the adoption of this advanced technique, whereas among museums there is still no common acceptance of the benefits that museums could derive from networks. In a library, when a catalog card is required for a new acquisition, it can be a duplicate of the same card required by every other library and, thus, can be printed, say, in Peoria from a computer file located anywhere in the country. With museum catalog records, it might occasionally be possible to produce a usable card (or another document) from a record in a file at a remote location, but in the majority of cases a museum object does not exactly duplicate any object that exists in any other museum, nor will cataloging in a central museum file adequately serve the information needs of other museums.

In this book, a good deal of emphasis will be placed upon the creation of information systems that are unique to the artifacts and information needs of each particular museum. Contrary to what was once believed, it is not feasible to develop one "ideal" cataloging system that will adequately serve a large number of museums and, in the process, allow the free and easy electronic interchange of all data about all objects in all the museums. The Canadian Heritage Information Network has demonstrated that a single system can serve an entire country, but even with this system it is still necessary for each institution to determine the information that it needs or wants to put into the system.

There are several reasons for this emphasis on creating individualized information systems for each institution (or organizational subdivision in the case of a large institution):

1. There are many different kinds of museums in terms of the kinds of objects that make up the collections. The information that is important, say, to an art museum, is very different from what is important to a history museum; and a natural history museum is just as different from both art and history museums as the latter are from each other. In addition to these general kinds of museums, however, are many other, more specialized, institutions. A photography museum such as the George Eastman House in Rochester, New York, for example, is in most ways similar to an art museum, but subject matter in the cataloging of photographs demands a precision that is unknown to most art historians. What is regarded as significant information in the recording of one category of museum object will always be different to some degree from what is significant for other categories.

2. Even when museums collect and exhibit identical or similar objects, the emphasis placed upon those objects—and, in turn, the information that is considered important about those objects—may be quite different. The best example of this is probably the archeological museum and the art museum. In the archeological museum, collections and exhibitions emphasize primarily the historical chronology and functional utility of the objects against a background of what is known about the society that made and used them. In the art museum it is the aesthetic importance and craftsmanship that are most often highlighted.

3. Prior to the era of modern industrialization, every human artifact and every specimen of nature was essentially unique. While preindustrial physical objects are always in some ways similar to other objects—it is this similarity that makes classification possible—they are never identical in the sense that machine-made nuts and bolts are identical to one another.

It is essential that we constantly maintain a clear distinction between the techniques of computerization, which can well be standardized across the bounds of most museums (the size of the collections being the major limiting factor), and the data fields and content of the information that any particular institution requires in order to fulfill its objectives and goals in the best manner possible. The latter are, in truth, almost unique in every museum because the objects themselves in each museum are almost unique. The more widely accepted the classification systems for the objects in a particular museum, the less need there is for creating new data structures. Geological specimens are probably described by classification categories and terminology that are more precise than for any other discipline. In a geology museum, therefore, the recording of specimens should be a very simple procedure, at least conceptually. At the other extreme, however, are art and photography museums, where subject access to visual resources creates all kinds of problems. The majority of the world's museums lie somewhere between these two extremes. Computers can help manage information, but the structure of the information that is important about collections must be determined separately for each institution.

2

Museum Activities—The Need for Artifact Records

Museum activities may conveniently be grouped, on the basis on when they are performed, into three major categories:

Initial Activities. When an artifact is first acquired, a number of things happen. The object must be accessioned, identified, registered, and possibly, restored. In some museums, a photograph is routinely taken as part of these initial activities.

Ongoing Activities. Once the first surge of activity is over, most museum objects rest in storage until such time as they are needed for a research project or exhibit, either at home or on loan in some other museum, or until a noticeable deterioration in the condition of the object indicates that some conservation effort is needed. It is these ongoing activities that are the essential business of the museum. Even though most of the artifacts or specimens will remain in storage, untouched for the life of the institution, appropriate groups must be readily locatable in order for the ongoing activities to be performed efficiently, and this is possible only if the initial activities were carried out properly to begin with.

Terminal Activity. Eventually, practically all museum objects are disposed of in one way or another. The terminal activity of decessioning (or deaccessioning) must be mentioned in order to make the cycle complete, but it is not a significant activity for this discussion.

The only reason for creating any sort of artifact record is to provide various individuals with information they will probably need to carry out certain defined activities—primarily information that is necessary or helpful to employees in the process of performing the ongoing activities of the museum. Ordinarily, research, exhibits, conservation, and loans cannot be based solely upon artifact records without physical examination of the objects themselves. Historically, however, card catalog files and the reliable (?) memories of good curators have been the only tools available to assist in locating the groups or artifacts that might be appropriate for any of these activities.

In order to lay the groundwork for what is to come, it is important to understand something about files in general, the kinds of files museums have maintained in the past, and the severe limitations the latter have imposed upon the ability of museum employees to efficiently perform their essential activities.

The word *file* comes from the Latin word for "thread" by way of the French word for "row." Thus it implies anything that is lined up in a series. Traditional examples include the pages of a ledger and cards in a file box, but the principle is the same whether we are looking at folders lined up in a filing cabinet, 3″ x 5″ typed cards to represent the artifacts in a museum collection, or an electronic file of the same information hidden from view within some kind of computer storage. The physical nature of these different kinds of files is distinctive, but their logical structure and content may be identical.

A file consists of from one to many *records,* with the information in each record representing all of the verbal (and sometimes visual) data available about one object or person. In museums, in addition to other records, there will almost always be a file of records containing available data on each artifact in the collection.

The structure of all records within a file should always be the same, i.e., each record should contain the same defined *fields,* or data categories, one for each type of discrete observation, kind of data, or category of information that will probably be required to record the objects that are the subject matter of that particular file.

Finally, the *content* of the data recorded in each field must be controlled in a consistent manner in all of the records contained in a particular file. There are several ways this can be accomplished, depending upon the nature of the data. For example, in recording dates, the structure of the field can be limited to a fixed number of numeric digits with punctuation to separate year, month, and day. In recording precise terminological fields such a object name or materials, the acceptable content can be controlled by word lists or thesauri, perhaps stored in computer memory. In like manner, the content of other fields can be controlled by various combinations of field length, alpha-numeric character limitations, syntax, and thesauri.

The important thing to remember at this stage is not the details, but the immutable structure of files: each file contains records with a fixed number of defined fields; and in each defined field the content is controlled in a precise manner.

The most common museum artifact file—probably the earliest in point of time—is what we now call an *accession file*. These files, of course, are not at all uniform. Some are maintained in massive ledgers, some are in less pretentious books, and some are simply file folders containing the paperwork (letters, signature sheets, etc.) associated with each acquisition that the museum has made. Generally, accession files contain *fields* of information such as the following:

- Accession Number—most often a two-part number corresponding to the year in which objects were acquired and the chronological number of their accession within that year
- Accession Date
- Source—name and address of the donor or vendor from whom objects were acquired
- Source of Funds—if the objects were purchased
- Method of Acquisition—donation/purchase/loan
- Number of Objects in the Lot

Notice that this file ordinarily does not include any object descriptions at all. If it did, a large bequest would require many accession records, and descriptions of each individual object would mean repeating a large quantity of accession information—which would not accomplish anything in terms of the activities of the museum.

Early on, curators and registrars conceived the idea of not repeating any more data than was necessary by creating a separate artifact *catalog file* in which the accession is referenced by the two-part number and each object is given a distinctive, third number (thus, 1925.23.7 would be the seventh object within accession number 23 during 1925). This innovation provided not only the possibility of separately describing each object without the redundancy of including accession data in each object record, but it also produced a convenient and brief system of code numbers that could be applied to the objects themselves as a permanent and positive form of identification.

Museum catalog files, like accession files, come in many forms. Some are little more than a card with the artifact number and a lengthy textual description of the object. Most provide for the separation of the textual description into a number of discrete fields of information. As one might expect, if the entire description is reduced to discrete fields or categories of data, the result is a catalog file with many distinct fields. A minimal catalog file, which can be maintained without too much work on 3″ x 5″ cards, might contain the following fields:

- Artifact Number–the three-part number described above
- Object Name
- Provenience–where did the object originate?
- Price/Value

- Artist/Creator
- Location–within the museum

Although most small museums are able to maintain both an accession file and a catalog file somewhat like those described, both files are usually organized in numerical sequence, which does nothing to aid employees in the performance of their work. The activities of the museum require the ability to find artifacts by such characteristics as object name, storage location, or artist name, not by a basically arbitrary number comparable to a social security number.

There are two basic organizations of itemized data. One can make a list of entities such as objects or persons, and attach a list of attributes to each entity, thus:

Artifact Number 1925.23.7
    Object Name
    Storage Location
    Etc.
Artifact Number 1925.23.8
    Object Name
    Storage Location
    Etc.
Etc.

Alternatively, one may list the attributes and attach to each one a list of entities:

Object Name ⟨1⟩
    Artifact Number
    Artifact Number
    Artifact Number
Object Name ⟨2⟩
    Artifact Number
    Etc.

——
——

Storage Location ⟨1⟩
    Artifact Number
    Etc.
Storage Location ⟨2⟩
    Artifact Number
    Etc.

The former organization seems natural to us as humans. We perceive a world of things, each of which exhibits certain attributes ("this *tree* is green"), not a world of attributes that show up in certain things ("*green* is manifest in this tree"). Therefore the first form of organization is referred to as *normal,* whereas the second is described as *inverted.* The inverted form of data organization, however, is invaluable for research. Indexes are the most familiar example, and finding artifacts appropriate to museum activities is, in fact, the same thing as a form of indexing.

In some of our larger museums the utility of inverted indexing was recognized early, if not by name, at least in practice. In the Henry Francis DuPont Winterthur Museum, for example, three copies of each artifact catalog card were prepared for many years. One was filed numerically, by artifact number; one was filed alphabetically, by object name; and the third was filed alphabetically, by provenience. This was clearly a great improvement over the maintenance of the artifact catalog solely in numeric sequence—at least the records on objects that had the same name could be located easily and it was possible to conveniently find records describing artifacts from the same locality. However, the system still had several drawbacks. It is possible to prepare three artifact cards without too much trouble, but when the number of discrete, indexable data categories is, say, ten or more, the clerical task of creating, filing, and controlling so many cards very quickly gets out of hand and the system breaks down completely.

Enter the computer and a totally new way of thinking about artifact records. No longer does one have to worry about the mechanics of what physically happens in the preparation and filing of cards. Once the information about each accession and each artifact is organized and entered into the computer files in the precise format demanded by the system, what happens physically, within the computer, is of no concern. The files can be reorganized on demand to fit almost any conceivable data need.

With the advent of data base management systems that can be used on relatively small and inexpensive computers, it is possible for the first time to concentrate initially on what kinds of records one wants to maintain: to begin—from the top down—with the objectives and activities of the institution; to outline a framework of records that will probably provide the greatest aid in carrying out those activities; to list the discrete types of data (the fields) that need to be present in each of the records; and finally, to organize the whole into a complete system that will provide the needed data to each employee in the most efficient and dependable manner possible.

Top-down organization such as this means, of course, that there must be a clear, communicated statement of institutional objectives and activities, and that someone must have a sufficiently organized mind to delineate the records and record fields that will be required in order for the total system to work. Although these needs appear to be obvious, in fact, failure to consider them at the beginning of the project is probably the biggest cause of problems with computer-based cataloging systems; success requires that the director and governing board must first do the hard thinking about their mission statement and the kinds of records that will be necessary to fulfill their mission most efficiently.

3

Computer Basics

This book is about the use of computers for a particular purpose within museums. It is not a book about computers as such, and we do not intend to discuss this modern-day marvel in any more detail than necessary. Nevertheless, it is essential that the reader have some basic understanding of what a computer is and how it functions. For the experienced "hacker," the first part of this chapter (prior to the "Data Control" section) can be skimmed over quickly or, if you wish, skipped entirely.

For those of us who have worked in universities or large governmental or business offices for as long as the last 20 years, it is hard not to think of rooms full of very expensive machinery that could be accessed only by preparing stacks of punched cards, placing them in exactly the correct order, and then leaving them for "processing"—whatever that meant—overnight. It is extremely difficult for us to accept the fact that many of the desktop computers that are available today are as powerful, in terms of internal capacity, as most of the room-size and expensive collections of equipment we were accustomed to using in the past. Our younger associates usually do not appreciate the humor when we tell them that we used to be able to tell what kind of a computer we were using by the shape of the little holes in the cards (there were only two: Remington-Rand equipment punched round holes, IBM created cards with rectangular holes).

Today, the three general classifications of computers on the basis of smaller to larger size are: microcomputers (the kind most of us have on our desks); minicomputers; and mainframe computers. As would be expected, the larger computers are a great deal more powerful than the smaller ones. With the more powerful computers, it is possible to *input* or *output* data from multiple, sometimes widely separated, geographic locations. The larger computers, of course, also permit the use of a wider variety of specialized input and output devices and they allow us to process extremely complex configurations such as colored pictures (see Part 3). It is essential, however, that we understand these differences as the result of magnitude or size only. Basically, all computers operate in the same manner and all computer systems, large or small, have the same general components: *hardware,* consisting mainly of thousands of on-off switches that are contained in those thumbnail-sized units we now call "chips"; *software* that instructs the proper switches to be open or closed at precisely the right moment; and *data* that are precisely edited, monitored, and interpreted so as to be understandable to a machine that is incapable of human thought.

HARDWARE COMPONENTS

The computer on which this is being written (a micro- or desktop computer) consists of only three major physical units: a nearly equal-sided, 10–12-inch cabinet with a small television screen and slot for inserting 3½-inch *floppy disks;* a separate keyboard not much different from a typewriter keyboard; and a printer. Inside of the cabinet, not accessible except to someone servicing the equipment, are several banks of chips that provide the necessary control circuits and one million *bytes* of random access memory (RAM).

Different brands of microcomputers have different features that are emphasized, just as Ford, Chrysler, and General Motors each has certain minor differences to give the salespeople some way of demonstrating the superiority of their products over the others. Some computers are able to emphasize economy by offering a package you can hook up to your own television. Others (in fact, many of them now) make room in the main cabinet for not one but two slots for inserting floppy disks. This makes it possible as you are entering data on the keyboard to store what you want to save on a "data" disk without the loss of time changing from the "program" disk. Some are wired to accept a *modem,* which will let one communicate with other computers via the telephone lines. Some are black-and-white only while others provide multicolor screens. And, of course, there are great differences in the size of the random access memory or in the combination of RAM and permanent, hard disks.

When one moves from the micros to the larger minicomputers and mainframe computers, the principles are the same. However, the hardware possibilities are much more numerous. As the processing needs of such systems expand, additional units of memory can be added to meet these needs (which is now possible, to a limited extent, with microcomputers as well). In addition, there are dozens of different types of specialized hardware, most of them equipment for entering and retrieving data and transmitting it to remote locations in more cost-efficient manner than is possible with just a keyboard and a printer.

Early computer operations were feasible for numeric processing, such as accounting operations, when the so-called computers were still electro-mechanical devices that could "read" and process punched cards. However, the complications of processing large alphabetic files demanded binary coding and true, electronic-speed processing that became possible only with semiconductors and high-speed chips. A brief review of the history of this development will perhaps assist in understanding why this is so.

The punched cards that for a long time were the only means of input into, first, accounting machines and, later, computers, were formatted so that each card contained 80 columns and each column had spaces for punching up to 12 holes. By convention, a hole punched in one or another column represented the numbers zero through nine in accordance with the decimal system of numbers. By punching a combination of two to six holes in a column, it was possible to record codes for the 26 letters of the alphabet and all necessary punctuation marks in addition to the numbers.

Now, if the first five columns of an 80-column card are defined, for example, as an arithmetic field representing check numbers, and the last ten columns on the card another arithmetic field representing check amount (ignoring for the moment what might be encoded into the other 65 columns), the functions that could be performed with a "deck" of these cards were:

> SORT—by mechanically reading the check number field, column by column from right to left, on a special machine called a card sorter, the cards could be placed in numerical, check number order
>
> ADD (one of several numeric functions)—and, at the same time
>
> PRINT—even with mechanical counters, each card could be read into the accounting machine and listed on a check register form, with the total of columns 71 through 80 accumulated in a counter and printed at the bottom of the list

This simple illustration, of course, is based upon letters and punctuation being coded into the cards in an essentially decimal-coded form (numbers were not coded at all). Today, the conversion of non-numerical data is not to a decimal, but to a binary-coded format, and the data are read initially into computer main storage rather than into cards. However, the functions performed are much the same:

> READ data that we convert into machine-readable, coded form by some type of input device, usually a keyboard
>
> SORT data *records* by *field(s)* into some predetermined hierarchy, by looking at each character in turn
>
> ADD (or perform a variety of other arithmetic functions), and

PRINT whatever results we may want to see (i.e., again converting data, this time from machine-readable, coded form back into a format that is comprehensible to the human intellect)

There are many additional refinements of these functions that are possible simply because we can now store much larger quantities of information on a temporary basis within a computer memory. For example, a very important function, practically unknown earlier, is the ability to compare the content of a field or fields in one record with that in another record within the file and to perform other functions based upon the result. In fact, the SORT function is now based upon a multiplicity of comparisons like this, performed very rapidly, with the results of each being to place one record either above or below another in the predetermined hierarchy.

All digital computer operations are based upon the principle of on-off electronic switches, and letters, numbers, and special characters are not simply translated into an electronic equivalent of the decimal holes on a punched card, but into a coded binary form (usually either EBCDIC or ASCII codes—see glossary). We may think of decimal circuitry as consisting of a series of switches, each with ten possible positions, zero through nine, whereas binary circuitry will have many more switches, but each one will have only two positions, zero or one. To anyone who is interested in comparing the two systems of notation, the glossary entry for *binary number* will be of interest. Any number, letter, or punctuation mark can be coded into two single decimal switches (remember, with cards each column could have multiple punches, with each column the equivalent of one switch), but with binary notation it requires four switches to handle the numbers zero through nine, and eight for all of the letters and punctuation.

In computer terminology, each binary switch is called a *bit*. Thus, when we hear that a computer is structured to handle an eight-bit *byte*, we know that it "bites off" or processes (e.g. reads, compares, sorts) eight bits at a time. Most existing microcomputers now have 16-bit processors; minicomputers use 32-bit processors and 64-bit processors are common with mainframe computers.

SOFTWARE

In general, hardware may be thought of as the computer equipment *per se*, whereas software is the totality of the programming necessary to make the equipment function in the desired manner. As with most generalizations, though, in practice this point of separation becomes somewhat fuzzy. With the computers of, say 40 years ago, it was true that a user had to start with what was then called *machine language*—the coding and control of binary switches in on-off positions. Within ten years or so after that, though, *assembly languages* had been developed that would permit a person to "talk" to the computer with precisely structured words and symbols that were translated into the actual binary, machine language necessary to make the hardware function. These were followed at various points in time by *programming languages* (e.g. FORTRAN, COBOL, RPG, PL/1, BASIC, and PASCAL) that made it more convenient to enter scientific formulas, business forms, and other instructions. Today, we do not even think about where our "natural" language breaks off and the binary, machine language begins. It is all hidden from view. The powerful programs contained on small disks instruct us to enter program parameters and data in English (or any other modern language) and our processing commands will be carried out. The distinction between hardware and software is still there. It takes control programs to make the hardware perform. In fact, even our printers have "computers" (i.e., computer chips) built into them so that they will function in a dependable way whenever the output from a particular user program is formatted in the manner prescribed as proper printer input. However, we now usually think of software as merely the multiplicity of user programs available to fulfill hundreds of special-purpose data needs.

The specific programming languages mentioned above are still the most common method of achieving one's objectives with a mainframe computer and a specialized staff of programmers to do the job. However, there are now available on floppy disks well over 1,000 programs to make a microcomputer perform almost any data task imaginable, from drawing charts and graphs to typing letters, doing accounting, and preparing income tax returns. The four most common categories of off-the-shelf programs are:

Word Processing

Almost every computer either comes with or has available some form of word processing program, sometimes a number of different word processing programs. These provide the user with a very convenient way of writing letters, memoranda, books, or any other document. The primary advantages over using a typewriter for the same purposes are the ease of erasing and the ability to move to another position in the same document or store for future use in other documents, anything from a single letter to whole paragraphs. Once something is "typed" into the computer memory it can be modified and reused as many times as necessary and in any way that may be desired, without having to retype the entire document.

All word processing programs provide some elements of style variation, such as italics, underlining, heavy outlines, shadowing, superscript, subscript, and often, the ability to change type size from as small as 9-point to as large as 24-point type.

In addition to style variations, some programs allow one to change at will from one font to another, e.g. Geneva, Helvetica, New York, Monaco, Chicago. It is also a great convenience, especially on a lengthy document, to set the desired format once and have it automatically carried throughout the document in a consistent manner. This includes the form and content in headings and footnotes, the numbering of pages, alignment (left, center, or right) and, if desired, justification of lines. Some word processing programs even go so far as to check one's spelling, raising a caution sign if a word is used that is not in the file of accepted terms so that the user must consciously override the standard spelling.

Most microcomputers can be justified solely on the basis of their use of word processing purposes. However, this is not the class of programs that is of greatest emphasis in this book.

Numeric Processing

In the introduction to this section, mention was made of programs to calculate income taxes. There are many of these, but they are only one small part of the group of programs that focus primarily on the completely dependable arithmetic or mathematical properties of computers. Accounting programs, whether written especially for a large institution or purchased on floppy disks, have been a major application since computers were first popularized. Others include: statistical program packages that allow a user to select from a number of business or scientific formulas what is necessary to calculate and print out a loan amortization schedule, depreciation calculation, or an internal rate of return; stock price calculations (often used in conjunction with a modem and the telephone lines to provide on-line or current price data); chart programs that allow one to enter data into particular cells on a spread sheet (preformatted on the screen) and calculate new information from the data entered and user-prepared formulas, display the new information in other cells, and print the entire chart; and many, many others.

Graphics

In addition to charts that, though graphic in appearance, are most often only a preformatted means of displaying statistical or other types of numeric data, there are available today a large number of programs for the production of various kinds of graphs, charts, and even custom art work on computers. When data are entered into or calculated by means of one of the chart programs mentioned in the preceding discussion of numeric processing, other programs can take the content of the several (or many) cells and display the same information visually in the form of bar graphs, pie charts, line charts, scatter charts, and other graphic displays. These charts can then be printed the same as any other information in memory, or combined in a variety of ways with textual data prepared separately using a word processing program. As one might imagine, this facility is another great aid to authors.

Another relatively new and rapidly expanding field is the production of multicolored artistic drawings by computer. Numerous methods are used, but usually some form of *cursor* or pointer allows the user to "paint" across the screen with a wide brush, a narrow brush, a brush with interrupted or dotted

lines, etc. When combined with the facilities of a powerful word processing system for the overlay of lettering, these types of programs have been a great boon to the advertising industry.

Data Base Management Systems

Data base management systems are yet another broad class of user programs, sometimes designed for a particular minicomputer of mainframe installation, but also available in the form of floppy disks for use with microcomputers. As with other off-the-shelf programs, each program varies in minor ways from all the others, and one must always be aware of the business-use bias built into most of them. However, data base management systems are program packages designed for the processing of *data files,* whether files of information on business inventories, employees, customers—or museum artifacts. From the discussion of files in the preceding chapter, it is easy to see the significance of this class of programs for museums. Undoubtedly the most crucial body of data in any museum is the files of information concerning the artifacts and specimens in the collections.

A data base management system consists of three different kinds of files, aside from data files *per se.* The most important are *program files,* which contain instructions in machine language setting limits upon various things such as field and record lengths and messages to be printed or displayed for the operator. *Control files* contain information for use by the system. For the most part they are managed automatically. Much of their content, though, originates with the user because some of these files record user decisions such as the names of files and fields and option choices that have been made. These two files may be thought of as a constitution, specifying all that may and may not be done and providing the user with mechanisms for determining exactly what is to be done, and when, within constitutional limits. As a rule, the overall system is regarded as immutable. Indeed, most packaged data base management systems have built in safeguards against modification.

A *directory file* is the system's record of where it put everything else. It is a file of files, connecting the user's file names with the system's physical file locations. It is used constantly. Often the user is allowed to view or print the directory, but the directory cannot be altered by the user. It is modified by the system as necessary when the user creates, extends, or deletes a file.

The four chapters in Part 2 are devoted entirely to data base management systems, with specific reference to the use of these systems by museums.

DATA CONTROL

In addition to hardware and software (the computer and the data base management system), the third element necessary for the creation of reliable and usable museum records is control of the data that is in the system. Data control should always emphasize the maximum availability of information in the files while at the same time enforcing whatever restrictions may be necessary to assure the accuracy or validity of the files. These objectives, on the surface, would appear to be in conflict with one another. However, with an adequate human-machine interface system both can be achieved at the same time.

With all types of manual files and with most of the older computer files, the entire system was predetermined and, with some exceptions, all information had to be collected and stored in one place. A manual file, of course, has to be searched manually. To some extent, it can be sorted to make it more usable for a particular need, but only at a cost of destroying the integrity of the original file. And the only way certain records or parts of records can be reproduced for use in individual, short-term projects is by copying what will be needed—a very laborious project in itself. With the older computer files it was possible to search, select, sort, and reproduce records. However, these were slow processes and they could be accomplished only within the structure of the fields and content organization envisioned by the creator of the system at the time it was first designed. Changing or adding to the structure later was extremely difficult and time-consuming.

With the computers and data base management systems available today a completely new strategy is possible. New files can be created as need dictates, often for purposes that were never envisioned when the system was designed and the first data entered. With commands such as SELECT, SEARCH, SORT, and COPY we can take from one file only the information that is significant to the purpose at hand, either certain records or certain fields within a limited number of records, and create new, probably temporary, files, sorting the records as necessary and adding to them any new data that may be significant.

As an example of the way this can work, let us assume that a museum has just received a new accession. The first procedure is to inventory the new artifacts, in the process determining the departments to which each will go and placing a temporary tag on each item with its assigned artifact number. As this is done, the accession/artifact number, and department identification, can be recorded and entered into a new computer file, perhaps named New Accessions.

The next step in the procedure is to create a new file for each department to use in further identifying, naming, and "registering" each object that has been received. These files may be thought of as temporary working files. As created, they contain only the department code, and the accession/artifact number. From the temporary files, worksheets can be printed with, perhaps, the department code and accession number in the heading on each page of the printout, and the artifact numbers, widely spaced, down the left margin of the pages. The rest of the sheets may contain only column headings, indicating the data fields to be filled in by each department curator.

When the artifacts have been described or registered by the various departments, the information can be entered from the worksheets into the temporary files, and again printed out for visual checking of the data. Once everyone is satisfied, the newly completed fields in the temporary files can be transferred into the permanent Artifact Catalog File. At this stage, the temporary working files can be erased, for any further use of the data should always originate by copying appropriate parts of the permanent Artifact Catalog.

With all ongoing museum activities, the same principle should apply: always start by copying from the permanent Artifact Catalog in order to be sure that new errors are not introduced by using superseded or otherwise inaccurate information. For example, when a new exhibit is being planned, information can be selected from appropriate Artifact Catalog fields: perhaps particular artists' names from the Artist or Maker field, along with all significant data associated with the artifacts of those artists or makers (e.g. date created, condition code). This information can be stored in a temporary file to be used throughout the planning of the exhibit, and printed as necessary for use as exhibit worksheets. When the exhibit planning is completed, the file can even be used (depending upon the completeness of the data) for the printing of the exhibition catalog.

The ease of copying files or parts of files for special purposes, plus the possibility of locating numerous small computers in different parts of the museum, suggest that many files rather than fewer will best accomplish the purposes of individuals and departments within the organization. And why not? Data freedom is one of the exciting things that is made possible by the present generation of computers. However, this freedom in the utilization of computer files raises two very serious questions: how much control over the data is necessary or desirable and how can this control be exercised?

Data control may be important at many different points within a museum, but three elements of control that *must* be maintained are:

Control of the System

In every museum there must be one person who knows and controls the system of accessioning and registering objects. This same individual should also be the one responsible for the maintenance of all permanent computer files, including the flow of data into and out of those files, although it is not essential that this person actually perform the physical operation of the computer. Usually, this central point of system control is the museum registrar.

Protection of the Master Files

It is also essential that a system be devised to protect permanent master files such as the Accession File and Artifact Catalog against unauthorized changes or deletions of records and accidental destruction by fire, flood, spurious magnetic fields, or other unanticipated damage. Several things can be done to achieve this protection:

1. Any time that changes in a master file are made (e.g. by the addition of new artifacts, changes of spelling, or even the correction of previously undetected errors), the entire file should be copied to provide a backup that is stored in some location other than where the master itself is normally kept.

2. Some data base management systems provide for password control over access to the files. What this means is that any time a file is accessed for any purpose, the user must enter an assigned, individual code before the file can be activated. Without a password code, the file cannot be entered, either for entry or retrieval of data. In addition, with most password systems it is possible to assign different codes to different levels of use—some codes can activate the files for retrieval of data, whereas other (presumably fewer) persons will have personal codes assigned that authorize the entry of data or the modification of existing data. Password codes, where available, are an excellent adjunct to prevent unauthorized or accidental changes in the files.

3. With systems that do not allow for the use of passwords, and perhaps even with those that do, it is always a good plan to maintain tight physical control over the master files. Where the means of storage of the files is on floppy disks, this can be as simple as keeping the disks themselves under lock and key and allowing them to be used only by persons who know the system and the limitations that exist over unauthorized changes.

Control over Input Accuracy

The third element of data control—sometimes the most difficult to achieve—is to take all steps possible to assure the consistency and accuracy of the data originally put into the master files. This will be discussed at various places in Part 2, but a listing of some of things that can be done to achieve input accuracy would certainly include:

1. Structuring the data fields, where possible, to minimize error. For example, on the *template* that is used to enter data, provide only the maximum field length allowed and pre-enter punctuation and "numeric data only" controls over fields such as dates, artifact numbers, and permanent storage locations.

2. With some data base management systems it is possible to include input vocabulary control, which tests the acceptability of words that are entered in particular fields against pre-entered lists and rejects words that are unacceptable or misspelled. This can be extremely helpful with fields such as Object Classification, Object Name, Style, or Materials, where the accurate retrieval of data depends completely upon knowing the vocabulary that was used when data was originally entered.

3. Where the system does not provide internal, computerized vocabulary control, it is always possible to prepare alphabetic listings of new entries in a particular field after the data have been entered (see *frequency dictionary* in the

glossary). This is not as good as catching the erroneous information before it is entered, but it does permit easy visual checking before the new entries have been copied into any permanent file.

Data control is a fact of life that must be recognized if one is to create dependable and usable master files. It is worth the effort, however, for with a few master files that are known to be completely accurate and dependable it is feasible to copy data as needed into numerous smaller working files. This, in turn leads to a system that gives the user the freedom to adapt

1. To the particular needs for which working files are created
2. To the changing data requirements that inevitably occur with the passing years
3. To the special research that is associated with different kinds of collections, e.g. the different data demands for collections of paintings or photographs as opposed to collections of furniture, toys and dolls, stuffed birds, or geological specimens

PART II

MORE ON DATA BASE MANAGEMENT SYSTEMS

4

What Is a Data Base?

The data base concept continues to evolve, but this much is certain: a data base comprises one or more computerized files serving a common purpose. These machine-readable files are analogous to such traditional ones as ledger books and boxes of cards. Their purpose is to hold information in the most easily accessible way possible.

In computer terminology, the word *file* refers to both physical and logical files. A *physical file* is not information, but only a container for information. It is comparable to a bound volume, which may hold a part of a book, an entire book, or two or three books. A *logical file* is a body of information comparable to a book. A physical file seldom holds more than one logical file, but a logical file may span several physical files. Physical files are divided into *physical records,* comparable to the pages of a book. Like pages, they are all the same size. In fact, they are sometimes called *pages.*

A logical file is divided into *logical records,* comparable to chapters or paragraphs in a book, each one devoted to a particular subject. Often each logical record has a physical record to itself, just as it might have its own card in a manual file. Many physical records thus contain empty spaces just as file cards do. In other files the logical records are independent of physical page boundaries, as though the items were printed in one continuous catalog.

Logical records are further divided into *fields,* with each field devoted to one aspect of the subject, such as (with museum artifact records) source or date of acquisition; and in some cases fields are further divided into subfields. Thus a location field might include subfields for room, unit, shelf, and date verified.

The person using a data base system should be concerned only with logical units, leaving the management of the physical storage to the system. In this book terms such as *file, record,* and *field,* refer to logical units unless indicated otherwise.

The logical structure of a file should be appropriate to the way it is intended to be used. There are unstructured *text files* (sometimes called *documents*) that are appropriate only for recording unstructured documents. There are also *structured text files, item files,* including rigidly structured *fixed format files,* and *relational data bases,* which contain fixed format files called *tables.* Each of these is discussed in the sections below.

FREE TEXT FILES

The simplest structure for a data base is none at all. Free text files are appropriate for certain kinds of documentary material, such as copies of deeds, for example. Their content is a running record of key strokes that represent letters, numerals, or other symbols and retrieval of such a free text file reproduces what would have appeared on paper if the keyboard had been on a typewriter.

Free text files cannot be used with some of the more sophisticated data base processes, but they can be indexed. Since an index file is structured, the content of a free text file is thus indirectly accessible for combination with data from other sources.

STRUCTURED TEXT FILES

Certain data base management systems use a type of structured text file in which each logical record describes one subject and is further divided into a series of paragraphs or *fields,* with each field describing one aspect of the subject, such as history or physical description. Since paragraphs are not easily sorted, each record of a structured text file may also contain a few short *sort fields,* with controlled content for recording indexable information such as date, object name, object number, and maker. Files of this kind are appropriate for descriptive catalogs, where the amount and nature of the data vary from record to record and natural language text plays an important role.

ITEMS FILES

Structured text files, strictly speaking, are item files since each logical record contains comparable data about one subject. Most item files, however, are more elaborately structured, on the model of a card file. Instead of having just two standard types of fields, these files usually include a variety of fields that are not only named but also described individually by the user in terms of data type, field length, syntax, and perhaps a variety of other restrictions. The user's field descriptions, called *data definitions,* are entered before any data is input and they govern the use of the file thereafter. DBMSs vary greatly in data definition options. In general, the more freedom or flexibility there is, the greater will be the costs of hardware, software, and operation.

Examples of flexibility in data definition include variable field length, unlimited length (i.e., free text), composite fields, unordered fields, optional, null, and repeatable fields, repeating groups of fields, and linked records. All of these concepts are defined in the glossary, and some of them are valuable in museum applications, especially for use with artifact catalog files. Variable length fields, for example, are desirable wherever proper names of people, places, or things are to be recorded. So are optional and null fields because so many kinds of data are essential for cataloging some kinds of objects but meaningless with other artifacts. Examples are fields for recording trademarks or paper type. Composite fields also have many uses, e.g. a measurements field, with subfields for data such as height, width, depth, and weight. Unordered fields can be useful when data are transcribed from a variety of sources that do not all have corresponding information in the same order; but they can also be dangerous, favoring accidental omissions.

Repeatable fields and repeating groups are essential if computerized records are to perpetuate traditional formats. A field for former owner (or ex-collection) may occur any number of times in a single catalog record. Ideally, it should contain subfields for collection name, place, year acquired, how acquired, and year of disposition. Also, in an accession file, where items are recorded by lots and each lot consists of from one to many artifacts, a repeating field or group is the best way to record the objects accessioned. In sum, repeating is the only efficient, reliable way to record *lists* of anything within a logical record. Warning: always view with suspicion any DBMS that allows repetitions but requires the user to set an upper limit upon their number. Such software may reserve space for the maximum number of repetitions whether the space is used or not. We shall return to this important problem under "List Processing and the Alternative" at the close of Chapter 5 and again in Chapter 11.

Linked records bind together objects that belong together in one way or another, as for example, the pieces of a table setting, matching sets of clothing or furniture, paintings from one altarpiece, or objects from the same accessioned lot. Typically, the links are *pointer* fields embedded within the records, so the system will retrieve records in related groups when that is desired. An object, of course, may belong to more than one group, as when a head and a torso represent the same sculpture though they were acquired in different accessions.

FIXED FORMAT FILES

Fixed format files are examples of the item file in its most inflexible form. They are the phone book model, supported by all DBMSs except those limited to free text processing as described above. In a fixed format file every logical record has exactly the same fields in the same order, though null fields may

Figure 1
Structure of a Flat File or Table

Number	Name	Room	Unit	Shelf	Date	Receipt
89.1						
89.2						
89.3						
(etc.)						
.....						

sometimes be permitted. Every field has a limit upon its length and none may be repeated within a record. The advantage of a fixed field format is simplicity, which translates to economy. Therefore, this type of DBMS should always be considered and should be used whenever it will not lead to a compromise of data or an unacceptable waste of space.

Fixed format files are perfectly acceptable for some types of collection records. Perhaps the best example is a location inventory file, where each record might contain seven and only seven fields: object number, object name, room, storage unit, shelf or drawer, date of verification, and receipt number (if on loan). Since location records must be kept up to date, changes may be frequent. The updating process is likely to be faster and easier with a fixed format file. Note, however, that if a cumulative list, tracking the movement of objects, is to be part of the recorded information, there must be at least one unrestricted repeating field or group, and the fixed format won't do.

FLAT FILES AND RELATIONS

Pure fixed format files are also called *flat files* or *tables* because their structure can best be visualized as a two-dimensional table in which each record is a row and each field or data category is a column (see Figure 1).

A relational data base is composed of similar tables called *relations,* in which columns represent fields and and records are rows or tuples (*sic,* as in quintuple, sextuple, etc.). So much for the mathematical mind at sea upon the English tongue! The objective of the relational model is to facilitate retrieval of any information that is either explicit or implicit in the entire data base, not merely to support certain foreseen queries. For illustration of this power see the example of *joining* in Chapter 7.

The relational data base separates information that might otherwise be combined in a single logical file into a series of relations. This analytical process, called *normalization,* takes place during the design of the data base. It is the

Figure 2
Published Reproductions by Artist and Title (un-normalized data illustrating data dependencies)

Accno.	Artist	Nat	Born	Title	Repro in	Auth	Pg
26.35	Beckman	Ger	1884	Family Picture	Ptg & Sculp	Barr	83
6.42.1-3	Beckman	Ger	1884	Departure	Ptg & Sculp	Barr	82
					Masters of..	Barr	63
246.56	Belling	Ger	1886	Sculpture	Bull.xxix,4	---	16
					German Art	Ritchie	174
230.48	Boccioni	Ital	1882	Develop...of a..	20thC It Art	Soby	12
						Barr	
520.41	Braque	Fren	1882	Table, The	Ptg & Sculp	Barr	103
					Geo Braque	Hope	109
						Cassou	

Figure 3
Data Dependencies in Figure 2

Column	Depends upon	and Indirectly upon
artist	accno	---
nat	artist	accno
born	artist	accno
title	accno	---
reproduced in	accno	---
author	reproduced in	accno
page	reproduced in + accno	---

Figure 4
A Set of Relations in the Third Normal Form

a. Published Reproductions

(primary key = composite of first and second columns)

accno	reproduced in	page
26.35	Ptg & Sculp	83
6.42.1-3	Ptg & Sculp	82
6.42.1-3	Masters of...	63
246.56	Bull.xxiv,4	16
246.56	German Art...	174
230.48	20thC It Art	12
250.41	Ptg & Sculp	103
250.41	Geo Braque	109

b. Works of Art

(primary key = first column only)

accno	artist	title
26.35	Beckmann	Family Picture
6.42.1-3	Beckmann	Departure
246.36	Belling	Sculpture
230.48	Boccioni	Devel...of a...
520.41	Braque	Table, The

c. Artists

(primary key = first column only)

name	nat	born
Beckmann	Ger	1884
Belling	Ger	1886
Boccioni	Ital	1882
Braque	Fren	1882

d. Publications by Author

(primary key = composite of first and second columns -- i.e., it is identical with entire row)

author	title
Barr	Masters of...
Barr	Ptg & Sculp
Barr	20thC It Art
Cassou	Geo Braque
Hope	Geo Braque
Ritchie	German Art
Soby	20thC It Art

exact opposite of such automated processes as selection and joining, in which data from the set of relations is retrieved, organized, and recombined into meaningful output. Figure 2 is an example of completely un-normalized data, with repeating groups and a complex system of relationships or dependencies between the various fields.

One column is said to *depend* upon another if a change to the latter could mean changing the former as well. For example, changing the attribution (artist's name) of the work entitled *Family Picture* in Figure 2 might also change the Nat and Born columns as well. Therefore, Nat and Born are said to be *dependent* upon Artist. A summary of these dependencies is shown in Figure 3.

For input and storage in a relational data base, information is broken down into a series of flat tables that satisfy the following rules, among others: (1) in each table one or more columns constitute a *primary* key, which uniquely identifies the subject of each row, and (2) every other column (if any) must *depend* directly upon the primary key. A set of tables obeying these rules is said to be in the *third normal form.* Figure 4 shows the data from our illustration, in the third normal form.

A detailed discussion of normal forms is beyond the scope of this book. It must be noted, though, that the higher normal forms (third and fourth) correspond to permanent data files in which the principle requirement is that every row describe one and only one thing. What this means for us is that an object description, for example, may contain an artist field because that tells something about the object being described, but it absolutely may not contain such fields as the artist's nationality or life dates because such information is about the person, not the object, and could change if the content of the artist field were different.

5

Data Base Organization

A data base contains many logical files, each serving an essential function and residing in one or more physical files. The logical files fall into three main groups: control files, data files, and temporary working files.

A primary objective of data base management is internal consistency. What this means is very simple: if a savings account shows a balance of $0.37, it must not show a balance of $0.58 at the same time. This would inevitably happen if the same datum—the balance—was recorded in more than one record, because only one can be updated at a time. Thus, consistency requires *non-redundancy,* a principle carrying so much weight in data base theory that an old definition was: ''A data base is a non-redundant collection of data relating to a major subject relevant to the enterprise'' (Inman 1986).

This is not quite acceptable today (as Inman recognized) because the data base concept has grown to include working files, which are redundant in their essence. But the non-redundancy principle holds for data files, the core of every data base, even though this principle runs counter to museum custom. It is usual with manual files, for instance, to find at least some descriptive data attached to accession and donor file records, in addition to more complete descriptions in the catalog. In the same way, catalog records often duplicate data found in the accession file and contain biographical data (*viz,* artists' birth and death dates) that are repeated in record after record.

Without computerization this practice made sense, for it was more useful, overall, to find the artist's biographical sketch on every object record than to be absolutely certain that each sketch was identical. This convenience, however, was balanced by the inconvenience of updating perhaps a few hundred records whenever an artist died or changed nationality.

KEY FIELDS

Keys are the strings and bolts that tie a data base together. They are data fields than can be indexed or used for sorting. All keys are classified as either primary or secondary. The difference is that a primary key must have a different (i.e., unique) value in each and every record of the file, while a secondary key may take the same value in any number of records. The importance of a primary key is that it forms a unique name of one and only one object. Thus, in a catalog file the unique object number in each record normally is also the number inscribed upon the object itself and the file number for all records, manual or computerized, that pertain to that object.

A key may be a single field or a group of related fields called a composite key. In such cases there may not be any single field that serves as a primary key but the conjunction of several may still provide the unique identification necessary to function as a composite primary key. For example, if a museum has included as a part of its object numbering system—in addition to year acquired, acquisition number and object number—a single digit prefix ("L," "D," or "P" to indicate whether the object is on loan, or was donated or purchased), the same value could be found in more than one record if this letter is not included as a part of the primary key. Yet this prefix also might want to be used as a separate sort field, say, to select all objects that are on loan. The best answer in this situation is to create four fields, each alone a secondary key with the combination of the four fields serving as the composition primary key. Thus, the concatenation of all four values (viz, P.1985.62.231) would provide the unique name of an object in the collection.

Any given key may appear in more than one file of the same data base and what is the primary key in one file may be a secondary key in another. The occurrence of the same key in multiple files, especially when it is primary in one and secondary in others, is the fundamental basis of relational data base theory and practice and is essential for such vital functions as *joining* files (see "Combining Files" in Chapter 7).

CONTROL FILES

As each new user file is set up, the DBMS requests information, usually by way of keyboard dialog with the user. First, the file must be named. Then a file organization must be selected if the system supports more than one. Examples are text files and item files. If an itemized organization is selected, then each field must be named and defined in terms of data type, length of field, and other options discussed previously. The permanent record of these decisions is a *data definition.* This may be entirely "local" (specific to one file) or wholly or in part *global,* applying to all files in a data base. Local components, including at least the assigned field names, become part of the new file and will disappear forever when that file is deleted. In many DBMSs, the entire set of field descriptions is considered to be local, having no effect upon any file except the one for which it is recorded. Then it is up to the user to make sure corresponding fields in different files of the same data base have the same description.

A *data dictionary* is a global set of field descriptions belonging to the data base as a whole and separate from any one file. In a data dictionary each field definition is named. Then, as data files are created, their fields are not defined locally, but only named and associated with a description in the data dictionary. The field name need not match that of the global description. The set of all fields throughout a DBMS that correspond to the same global description is called a *domain,* and the name of the description is also the name of the domain.

The advantage of a data dictionary is not the mere convenience of defining a field in a single word. It is that corresponding fields in different files are forced to share the same description, which avoids problems with such powerful capabilities as joining files (relations) in a relational DBMS. A data dictionary is one of the most desirable features to be sought in a DBMS.

An *authority file* is a list of acceptable values for a given field or, preferably, a domain. The automatic checking of input against an authority file prevents entry of misspelled and unauthorized terms, holding subsequent corrections to a minimum. Equally importantly, it draws attention to first-time entries. Thus, if a new artist's name is rejected by the system as unrecognizable, then some authorized person will have to designate a standard spelling, update the authority file and, if appropriate, create a new entry in a separate file of artists' biographical data. An authority file is also a useful guide to data retrieval. If you examine the authority file first, you know what terms may be sought in a *query* and exactly how they are spelled, capitalized, and punctuated.

Depending upon the nature of the domain, an authority file may be as short as "Y" or "N" in a *flag field* or as long as an open-ended list of persons, places, or things. There must be a way of adding to an authority file, since lists of proper nouns can never be complete, but the authority to update such a list must be reserved to one or very few competent persons whom the system recognizes by password.

A few DBMSs do not store in memory the terms that are input but only short codes in the form of binary integers. This requires an authority file that includes all the recognized terminology plus a code for each. To be useful, code numbers should be assigned in alphabetical order, so they will sort in the same way as the terms they represent. Sometimes variant forms of a term, such as "Rembrandt" and "Van Rijn, Rembrandt H." may be accepted as input but assigned the same code. Upon decoding for output, though, one preferred form will always be used.

Coding has the disadvantages that all input and queries must be coded (which, however, is no more difficult than checking an authority file) and all output must be decoded. In addition, great care is necessary in assigning codes. The original objective of coding was to save storage space, which was once very costly. However, there are other advantages. All codes are of the same length and data type and, on the average, are much shorter than the terms they represent. This translates into greater efficiency in all processing except output. There is no shortage of numbers. A four-byte integer, for example, could be assigned to every entry in *Webster's Unabridged Dictionary* and there would still be about 10,000 unused numbers available for assignment to new words between *each* successive pair of entries. Coding is suited to large bodies of long-lived, rigorously controlled data, especially in archival applications where the majority of items are retrieved infrequently.

Thesauri are like authority files except that each defined term may also be associated with one or more subcategories maintained for preferred terms, synonyms, narrower terms, broader terms, and related terms. A thesaurus may be manual or automatic and, like an authority file, is used to control input vocabulary and as an aid in framing queries. In particular, thesauri permit the expansion of queries as in the following example:

FIND (medium = silk screen) . . . might be expanded to
FIND ((medium = silk screen) OR (medium = serigraph)).

Automatic query expansion can be dangerous without the strictest control of vocabulary. Imagine a query for "wood" expanded to include all possible narrower terms such as "oak," "pine," and the multitude of subspecies. The mere process of query expansion would swamp many systems. Then, too, many synonyms are synonyms only in one context. A serigraph is not only a silk screen print but also a machine for gauging the strength of raw silk.

Many DBMSs support the storage of complete procedures or *macros* that are used frequently. These are a series of commands used to accomplish some repeated task such as generating a monthly report of accessions and decessions. Instead of entering the entire series of commands at the keyboard every month—and probably making mistakes—the user issues a single command, which is the name of a file that contains the entire series of commands.

DATA MASTER FILES

Data files are the heart of every data base and normally the bulk as well. They are also known as *master files* and *truth files,* and their content as *primitive data.* They tend to be long-lived, often outlasting hardware and software systems. They are required to be consistent—hence, non-redundant—and up-to-date as far as resources permit. Examples are accession files, inventory files, and catalog files. Data files about persons, organizations, literature, and events

Figure 5
Characteristics of Primitive and Derived Data

	Primitive Data	Derived Data
Source	Input	Other files
Redundancy	Undesirable	Unavoidable
Consistency	Global	Local
Duration	Permanent	Temporary
Currency	Update as possible	Re-generate if necessary
Habitat	Data files	Working files
Purpose	Storage	Research

may be public in nature since many institutions need the same information. In these areas a museum may be a user of data files compiled and made available by others.

The difference between data *master files* and subsequent *working files* is that the former contain primitive data that came from the outside world by way of a keyboard, optical scanner, or automatic sensing device, whereas the latter contain *derived data* that came from primitive data by way of selection, copying, and computation (see Figure 5).

The reason for these distinctions is that derived data can only reflect the state of primitive data at the moment of derivation. Thereafter, they are separate and disconnected. Updates of the primitive data files do not automatically affect the derived data files, and the latter, through the passage of time, become more and more inconsistent with the files from which they came.

Museums keep certain permanent data files that contain at least some derived data. With or without automation, these files are a perennial cause of consistency problems. Typical examples are *source files* such as a donors file or a vendors file. Some of their content comes, directly or indirectly, from the accession file. Other parts, such as addresses, phone numbers, and comments, are not—or should not be—in the accession file because they would have to be repeated in so many records. Consistency problems can be minimized by limiting duplication of data to the single key field needed to relate the files—in this case, the name of the source, where the Source Name field in the accession file and the Name field in the source, donor, or vendor file belong to the same domain and serve to relate the two files (see Figure 6).

Figure 6
Relating Files by Means of a Common Key

Accession File	Source, Donor or Vendor File
Accession number*	Name*
Source name	Address
Date	Phone number
Method of Acquisition	Comments
(Other fields)	

"Source Name" in the Accession File and "Name" in the Source, Donor or Vendor File belong to the same domain and serve to relate the two files; * = Primary key field

WORKING FILES

Working files support research. As compared to data files, they are further from input and closer to output because they contain derived data as discussed in the preceding section. A working file begins with selection of data in support of a specific research goal, research being understood to include everything from looking up telephone numbers to compiling an encyclopedia. Anything added by direct input to a working file must be of temporary interest, relevant to the research but not to the permanent records of the institution.

A working file is temporary. Strictly speaking, "temporary" can mean anything short of eternal, but the sense here is that the disconnection of derived data from permanent data files means that the former inevitably go out of date. This is unimportant for short-term research (the great majority of projects), but a long-term project that has been interrupted should always recommence with a fresh selection of working data.

Not every working file is derived directly from permanent data. Many, perhaps most, are derived from other working files. The case study in Chapter 6 demonstrates what usually happens.

LIST PROCESSING AND THE ALTERNATIVE

As we saw in Chapter 4, the ability to record and retrieve repeating groups and repeatable fields is vital in museum data base applications. A museum's principal files are all *catalogs* of objects, persons, places, events, and concepts; and in a catalog repeating groups and fields are unavoidable. Objects consist of multiple materials, have multiple parts, come from a series of former collections, have been in multiple exhibitions, and are cited in multiple publications. Accessioned lots include multiple objects. Persons and places are known by multiple names and events, such as exhibitions, involve multiple objects and any number of persons.

The ability to store and process lists, i.e., repeating groups and repeatable fields, is called *list processing,* a term that is also applied to a number of other data processing concepts (see glossary). DBMSs that support list processing are, however, in the minority. The reason is simple: this facility adds a little time to the processing of each file transaction, hence a little cost. In business applications, these tiny costs per transaction add up and it is often more economical to design records without repetition than to pay for list processing.

This leaves the museum community—and most of the scholarly world—in a dilemma. What can we do when our data demand repeating fields and our software forbids them?

We begin with the worst solution, mentioned here only because it is often the first to occur to someone facing this problem for the first time. The worst solution, then, is to design a record format with more than one field for the same category of data, for example, the fields Excol1, Excol2, Excol3, and Excol4, all for the names of former owners. The effect is to complicate retrieval enormously, or preclude it together, because any ex-collection you look for may be recorded in any one of several fields. In addition, space is wasted since most objects won't have four ex-collections; and, no matter how many Excol fields you may include, there will never be enough for the exceptional case.

A better solution, if the DBMS supports long text fields, is to build a list of data within a single, variable-length field such as:

> Materials: oil and newspaper collage on canvas over plywood with charcoal and pencil

This will usually accommodate the data, but selective retrieval must depend upon either sequential search, word indexing of the text field, or, in the best of cases, indexing of text phrases. These ideas are discussed at length in the following chapter.

With a relational DBMS and the *joining* facility there is an elegant, general solution to the problem of multiple values. It involves a rather unfamiliar organization of input and storage which, however, is characteristic of relational DBMSs. We shall examine how it might be applied in the case of an object catalog listing all known ex-collections (former owners) for each item.

In this relational organization of files the format of the object record is normal except that all the data that could be repeatable are omitted. What remains is something like this:

- object number (primary key)
- object name

- measurements
- weight
- department
- etc.

There is no reference at all to former owners because that is repeatable data. At most there might be a ''counter'' field containing the number of known ex-collections. Then a completely separate ex-collections file is set up. Its records have only two essential fields and the primary key is the composite of both. Thus:

Column 1	*and*	*Column 2*
object number		collection name (composite primary key)

Each record in this file associates one object, by number, with one uniquely identified collection; and although each record refers to only one object and one collection, the file as a whole may associate any number of objects with each collection and vice versa.

The separation of the two files is, of course, only for storage purposes. A system that obliges one to keep multiple files of this kind must offer the facility to recombine the data for readable output, as discussed under ''Reports'' in Chapter 7.

There is no reason why the ex-collection record format may not be expanded to include additional data such as year of acquisition, source of acquisition, method of disposition, or year of disposition so long as every field refers only to a single interlude of ownership of a single object. Thus, the complete ex-collection file might provide for the following fields:

- object number) the composite
- collection name) primary key
- year of acquisition
- method of acquisition
- source of acquisition
- object number in the collection
- year of disposition
- method of disposition
- next owner (possibly your institution)

The record may not include data about the collection *per se* (except its name or other identification) because that would be redundant from one record to another; and, needless to say, it may not include repeatable fields such as bibliographic references. If the collections as such are cataloged, they form yet another file in which the domain Collection Name is the primary key.

Finally, please bear in mind that this section is not specifically about recording ex-collections but rather the fundamental problem of getting along without list processing. The technique illustrated here for ex-collections applies in all similar cases, another instance of which is discussed in Chapter 11.

6

Finding Data

The problem is to prepare a list of the dolls in a collection that were made in North America during the Victorian period and are available for use in a traveling exhibition. The permanent data file is a museum object catalog that contains the following fields, among others: Object Number, Object Name, Country of Origin, Earliest Possible Date, Latest Possible Date, and Loan Status (a flag field indicating whether or not the object is available for loan). We assume perfect vocabulary control, so that the records representing every doll in the collection have the word ''doll'' in the Object Name field.

The first step is to derive a subset or universe of data very much smaller than the whole catalog. A query to accomplish this might be

FIND (object name = 'doll')

to which the system would respond by displaying on the screen

OBJECT NAME = 'doll' 000528 RECORDS FOUND

This being a satisfactory number, the user would issue a STORE or SAVE command that would result in the creation of a new 528-record working file derived from the catalog, which need not be used again. In physical storage, this working file might actually contain all of the 528 records; in some systems it might hold nothing but a list of record numbers and *pointers* referring to the location of the records in the permanent catalog. The difference would not be visible to the user. Either way, this is a new working catalog, and there are still a number of queries to be made in order to narrow the field of search. Which question to ask next is a matter of guesswork and experience with the file. A good choice might be the nationality criterion. If the system supports *boolean* queries (most of them do today), the query would be

FIND ((country of origin = 'Canada') OR (country of origin = 'US'))

Both parts of this query would be performed with a single search of the 528-record 'doll' working file. If the system does not support boolean searches, however, it would be necessary to question the file in two installments, with both queries addressed to the same 528-record file. If the second part of the question is addressed to a new file, created after the first search, the second search would not add any new records.

The next query might use the period criterion:

FIND ((earliest possible date > 1836) AND (latest possible date < 1902))

This is a boolean AND query, so the two halves need not be addressed to the same file if they are asked separately. It would be possible to extract a file of dolls not earlier than 1837 and then extract from the resulting file a shorter list of those not later than 1901.

There is only one criterion left:

FIND (loan status = 'yes')

The resulting set—if anything survives—is a working catalog of North American dolls of the Victorian period that are available for use outside the museum. If all has gone well, the intermediate working files may be deleted.

Extracting records from data files is only one of many processes used to derive working files. Other techniques include combining files, sorting records, extracting individual fields, and counting and statistical manipulation of numerical values. No matter what combination of processes may be employed, the objective is a step-by-step progression from general purpose data storage toward a specific research goal, the results of each step being saved in working files for as long as needed.

SEARCH STRATEGIES

The search illustrated above utilizes queries without considering how the system is to carry them out. In fact, there are a number of search strategies, some of which perform better than others for certain kinds of queries; and some systems would have found it difficult or impossible to handle some of the queries in the illustration.

A query is a *command* that instigates a search operation. Like any command it begins with a *command word* known to the system, followed by a *parameter list* that provides whatever information may be necessary to carry out the command. In our illustration, the key command word was FIND. To carry out a search, the system had to know two things: where to look and what to look for. We assumed that the question of where to look had been settled by some earlier command such as

```
OPEN FILE (CATALOG1)
```

Otherwise the FIND command would have needed a two-part parameter list such as

```
FIND (CATALOG1) (object name = 'doll')
```

Here CATALOG1 is the name of a logical file. It is common to search two or more logical files at once, provided they are organized the same way. Thus the command might be

```
FIND(CATALOG1,CATALOG2,CATALOG3) (object name = 'doll')
```

The parameter *object name = 'doll'* constitutes a simple, matching query. It tells what to look for: in this case, records in which the value of the Object Name field is exactly the character string *doll*. Often, though not always, the equal sign is used to indicate such an exact match. Other search criteria supported by some systems and indicated by other symbols include the substring match (the field contains the character string *doll*), the leading substring match (the field begins with the substring, *doll*), and various others. Comparison operators such as equal to, less than, and greater than are used primarily with numerical fields, but also with character fields, where *less than* (<) is interpreted as earlier in alphabetical order, and so forth.

A class of query called *range queries* is extremely important in collection-based research. Some search strategies handle range queries poorly or not at all, and systems with this failing are useless to us. A range query may be expressed as

```
FIND (date = (1837–1901))
```

or, in other systems, using the boolean form:

```
FIND ((date > 1836) AND (date < 1902))
```

Either of these would be satisfied by any record where the value of an integer in the Date field is at least 1837 but not more than 1901.

Boolean queries use parentheses exactly as in algebra, plus the three boolean operators AND, OR, and NOT. The NOT operator simply reverses the sense of whatever follows in parentheses.

Sequential Searching

One way to satisfy search criteria is to go through the entire file or set of files evaluating each logical record in turn. This is called a *sequential search.* It is the slowest of all strategies and therefore unacceptable for large files, unless time means nothing. It is also the easiest search to program and the least demanding since it requires neither indexing nor sorting. A sequential search handles every kind of query. It is the only way to search a file stored on tape and the only way to handle a substring match or to find anything in a text file without word indexing. The time required for a sequential search on a simple query is proportional to the number of records searched.

Binary Searching

This strategy works by closing in upon its objective as people do with a dictionary or a telephone book. It looks near the middle of the file and, depending upon whether the value found there is too high or too low, it then goes forward or backward a quarter of the length of the file, and so on, back and forth by smaller and smaller increments until it reaches the value sought—or the place that value would occupy if present.

A binary search requires two things: first, the file(s) must be on some type of direct access device—usually, this means a disk—and second, the file(s) must be sorted on the field that is to be queried. It follows that no query may refer to more than one data field. For example, the range query

FIND (date = (1837–1901))

would work in a binary search, while the boolean range query

FIND ((earliest possible date > 1836) AND (latest possible date < 1902))

would not work because a file cannot be sorted on two fields at once. If the file is in order by earliest possible date, then it cannot at the same time be in order by latest possible date. This would require the file to be resorted after the first part of the query had been processed; and sorting takes even longer than a sequential search. If your DBMS requires sorting before a query and does not support boolean queries, then it undoubtedly uses the binary search strategy.

For files of more than ten records a binary search is much faster than a sequential search. Ten file accesses suffice to search a thousand records, and 20 accesses can span a million. However, a search that involves ten or 20 file accesses is considered rather slow.

Indexing

The search strategies discussed above all imply that the content of particular fields within a data base are examined to determine those records which meet desired criteria. For some kinds of data—generally, numeric data fields or flag fields—these strategies work well. However, with alphabetic data fields such as object name, artist name, style name, or materials, it is sometimes more efficient to prepare indexes to field content and to search only the indexes for possible matches rather than to search the entire data file, which probably will not be sequenced according to the content of the field being searched. It should be recognized, though, that indexing is a type of search strategy.

The word *index* comes from the Latin verb meaning "to point out"—the *index* finger is the one used for pointing. An index is a file, separate from the indexed data file or working file. Like a book index, it consists of an ordered list of things to look up, each associated with a list of one or more location *pointers* comparable to page numbers.

Software systems vary greatly in their use of index files. To understand this one must bear in mind a number of points. First, all indexes consist of data *derived* from the file(s) indexed. Therefore, an index, left to itself, will be out of date the first time the file from which it was derived is modified, and thus will no longer correspond to what it is supposed to index. Some systems rely upon temporary indexes, generated when they are needed and then deleted. Others use permanent indexes that are updated automatically whenever the data files change. The latter is a very complicated process.

Second, index generation is time consuming, much more so for long files than short ones. If a system can index a 1,000-record file in one minute, a 2,000-record file will take more than two minutes and a 20,000-record file very much longer than 20 minutes. Short indexes are expendable, while long ones represent a fairly large capital investment. Systems designed for short files, therefore, tend to rely upon temporary indexing, while those intended for massive data bases support automatic index maintenance.

Third, text files and free text fields are indexed differently from other data types. For text files, *word indexing* is essential. Words are indexed not only to the file, record, and field but also to the paragraph, sentence, and word position. Often this is the only way to locate a name or a phrase of more than one word. Word indexing entails queries such as

```
FIND ('United' followed by 'States')
```

that could not be answered without knowing the exact position of each word.

Because indexing every word is useless and expensive, word indexing is usually controlled either by a *stop list* of articles, conjunctions, prepositions, pronouns, and auxilliary verbs that are not to be indexed or, alternatively, by a list of *keywords* (or phrases) that are the only terms to be indexed. Unless it is limited to a short list of keywords, a word index may end up as large as the file being indexed.

Field indexing is used for data types other than text. Typical queries based upon whole field contents or values have been illustrated in the preceding section. A field index is about 1–15% of the size of the indexed file, multiplied by the number of fields indexed.

Fourth, many systems, especially those with relational ambitions, generate and maintain indexes automatically. What the user observes is (1) queries handled as if by magic, and (2) extra files appearing in the disk directory in addition to those created by the user (or else physical files much longer than the number of stored bytes would suggest).

Fifth, some so-called indexes are not indexes at all, for they do not point. Instead of location pointers, these files incorporate the entire content of every record indexed. This avoids having to access the data file after an index search, but it means that every index is about as long as the file indexed—often considerably longer. Thus, a data file indexed on only five fields might require an index file six to nine times as long as the original file. There is nothing wrong with this technique, other than the dishonesty of calling it indexing. It is used primarily by microcomputer DBMSs such as Dbase III(TM), where files are expected to be short and indexes temporary.

There are many strategies for searching an index, each entailing a specific file organization. The majority use a multi-level *tree* structure. The search begins with the root segment of the term for which a match is desired—perhaps two hundred root segments being distributed evenly through the index. If the target term is not a root segment, then it must fall between two of the terms that are there and this directs the search to one of the many segments at the next level. A *hashing*-based index relies upon different principles. Instead of a step-by-step examination of the index, the system calculates the location of its target and then goes directly to it. However, there are some problems in this, which sometimes result in longer searches and/or a considerable waste of space, especially with data such as that contained in a museum catalog. If your system does not perform range queries or has difficulty with long ranges (e.g. 1837–1901) and open-ended ranges (e.g. < 1901), then it relies upon hashing exclusively and you have serious trouble.

Many DBMSs use not one but several search strategies, selecting one method or another depending upon the nature of the query. INGRES(TM), for example, uses hashing where appropriate but switches to another strategy for range queries.

7

Using Your Data Base

All DBMSs work in close cooperation with the *operating system,* the master program that controls and monitors all operations of the computer, depending upon it for many essential services. In some cases the DBMS is, in fact, an integral part of the operating system. More often, though, it is a separate software package, designed to cooperate with a particular operating system. Some DBMSs exist in several versions, each compatible with one operating system or another.

Compatibility with an operating system means that the DBMS frames its requests for service in the command language of the operating system. It also means that data passed to the operating system, as for storage or screen display, goes to the storage location where the operating system will look for it and is in the format the operating system expects—and vice versa—i.e., the DBMS knows where to look for data from the operating system and how to interpret its format.

The operating system controls password security, if such is a part of the system, allocates physical storage, and records physical file locations in the directories and tables of contents, including storage of installed software such as the DBMS itself. The correspondence between the DBMS's logical names for files and the operating system's records of physical locations is a critical aspect of communication between the two systems.

Most software systems, DBMSs included, have a mechanism for calling upon *applications programs* for special services that are not built into the software in question or the operating system. Some applications programs are off-the-shelf packages while others may be written by or for a particular user or group of users. Usually an applications program can be written in any programming language, but it must be compatible with the DBMS in the same way a DBMS is compatible with its operating system.

Typical uses of applications programs include *preprocessing* and *postprocessing*. Examples of the former are programs that check spelling, convert measurements from one system to another (e.g. English to metric), or derive ratios between one field value and another. Postprocessing may include the same functions as well as statistical analyses, checking the format of a report, inserting diacritical marks, and the preparation of visual displays such as maps and graphs.

Every machine made has built-in limitations. An automobile can go just so fast and carry just so many people. Such limits can be exceeded only at a cost in time. Thus, a five-passenger car, able to carry seven in a pinch, can move up to fourteen people from one place to another, but only by making two uncomfortable trips. Computer systems have analogous limitations, some built into the hardware, some imposed by operating system design, and some added by your DBMS for the sake of economy and simplicity. These inherent limitations vary from system to system, but they always represent a trade-off between processing capabilities, storage space, and time, all of which cost money.

DATA TYPES

Users feel the impact of build-in limitations in many ways. One of the most important is the selection of *data types,* which affect what you can store and process, your ability to communicate with external systems (such as those of other museums), and what may be most important of all, your own future system, which will someday inherit the data.

As we know, digital data are represented by a series of bits that have no meaning in and of themselves. A data type is simply a convention for representing an *idea* by a certain sequence of consecutive bits. The more complex the idea the more bits it requires. Data types can represent any kind of idea that can be defined in finite terms. The *character* or *text* type is basic (almost all systems now use either ASCII or EBCDIC standard character codes) and every system has some standard method of recording characters internally in binary form. All keyboards transmit character data and all printers require it. Hence, it is the primary mode for input and output. *Binary numbers* are also basic because computers can only do binary arithmetic. Most DBMSs support one or more numeric types, though some conceal their binary operations, pretending to do arithmetic with character data, just as people do. Numeric data types include fixed and floating-point binary numbers, fixed and floating-point decimal numbers, long (four-byte) and short (two-byte) integers, imaginary, and complex numbers.

Most DBMSs support *flag fields,* though in some the user must represent yes-or-no data by the characters "Y" or "N" or the numbers zero or one.

Common "special" data types are dollar fields, time fields, and date fields; however, beware! The majority of DBMSs are designed for small businesses. The date fields required for business use will not answer the needs of museums—they cannot represent a date that isn't exactly known to the day (i.e., they cannot represent a date expressed as an era, an approximation, or an encompassing range of years); they cannot represent a date more than a few years in the past; and often they cannot represent a date in a manner that can be communicated to another system.

Data types are not universal. Even the commonest types, such as text, are defined and used differently and called by different names in different systems. Therefore, when evaluating a DBMS, the potential user cannot assume anything from the names of data types but must understand the rules that define each type for that system. Only then is it possible to know whether a system can support a planned application, whether data already stored in another system can be carried over, and how difficult it may be to get the data out of the system when it is outgrown or obsolete. One rule of thumb is to avoid special data types, which are seldom transportable from one DBMS to another.

CONTENT CONTROLS

Many systems allow a user to impose additional constraints upon the content of individual domains and fields, over and above those implied by the choice of a data type. The purpose of such constraints is to catch as many input errors as possible at the time of entry. Users' rules are a part of data definition and, once declared, are enforced automatically. Field content controls may be immutable, which is dangerous; more sophisticated systems merely warn the user in case of "illegal" input. The warning can be overridden. Typical user-imposed content controls are

- a one-character field taking only "Y" or "N"
- a Year of Origin field that accepts no value higher than +2000
- a Place Name field that takes no character that is not a letter, a period, or hyphen
- a field that will accept only terms found in an authority file
- an Object Number field that accepts only the characters that are used in the museum's object numbering system

DIALOG WITH THE USER

On-line DBMSs carry on a conversation with the user, requesting instructions, listing and explaining possible courses of action, pointing out problems, and reporting activity. Some require typed command statements learned from a manual. Others present *menus* of possible commands, from which the user can select by typing a number or moving a *cursor* on the screen. Some work either way. Typed commands are generally easiest for the experienced user, menus for the inexperienced.

A major aspect of dialog with the user concerns error handling: the detection and reporting of abnormal conditions, including not only catastrophic events such as power failure and hardware breakdown but also program "bugs" and data errors. Software can do little or nothing about mechanical and electrical failure, and not much to diagnose its own faults. At best, a DBMS that has taken a wrong turn may sometimes have enough sense to terminate its own run and return control to the operating system.

A well-designed DBMS should never be at a loss, and nothing the user can do at the keyboard should confuse it. The proper response to a data error is a clear message on the screen, explaining exactly what has gone wrong and requesting instruction. Among options offered should always be the QUIT and HELP commands.

INPUT

One of the primary functions of an on-line DBMS is to control and facilitate input to data files, once the files have been created and the record format defined. The objectives are to make data capture as fast and easy as possible and to minimize the need for revision by catching as many mistakes as possible at the time of input.

As a rule, input proceeds record-by-record, and new records do not go directly to a permanent file but first to a temporary holding area, either in main storage or in a temporary disk file. Then they may be recalled and reworked until satisfactory, deleted (after practice input, for example), or copied into permanent files by a STORE command.

Input to a text record or field usually begins with the cursor at the upper left corner of a blank space. The user has some or all of the following options:

- skip the field, leaving it empty or null
- insert a copy of text stored elsewhere for repetitive use
- insert a copy of the corresponding field from the previous record
- type input

The first three of these options often require no more than one or two key strokes.

The typing process is assisted further by an array of text editing facilities such as

- erasure by backspacing
- moving the cursor within text already typed to make corrections, insertions, and deletions

Figure 7
An Accession File Template

```
Wed. Feb 22. 1989 / 11:23:44am / PROC = APPEND / FILE = ACCS

accno__________

date __/__/__

source ____________________________

___________________

fund source ________________________

___________________

mod of acq __________

count _____

remarks
```

The rectangle surrounded by rays represents the bright cursor. No space is marked out for the free text "remarks" field, which is presumed in this illustration to have unlimited length.

- inserting at any point a copy of text stored elsewhere
- many, many other editing commands

If the system includes a spelling checker, each word may be validated as it is typed, with unknown terms evoking a warning bell. In other systems, the user must call for validation after the text is complete. This produces a list of unrecognized words.

Item file input is often assisted by a *template,* a blank form projected on the screen. If the record format is longer than one screen, only the top may appear at first, the template "scrolling" upward as lines are filled. If the format uses only half the screen or less, the previous record may show at the top, for reference, with the current template below. In the template, each field is represented by the field name at the left margin (or, sometimes, at the top of a column), followed by punctuation, then a space to be filled. Usually each space is marked out by underscore characters indicating how many character positions are available, though unlimited text fields are an exception (see Figure 7).

The template need not be empty initially. Fields may appear with a copy of the corresponding field in the previous record. This can be very efficient where consecutive entries are similar, as in cataloging a collection of photographs all from the same accession, made by the same photographer in the same year in the same city, and printed in the same format on the same paper. Fields may also appear filled with default values such as "unknown," "NA," or "Yes" in cases where such entries are frequent.

Each new record template appears with the cursor at the beginning of the first data field. The user may fill the field and, when all the space is used or a carriage return key or tab key is typed, the cursor will skip to the beginning of the next field. In many DBMSs, though, there are options other than typing the field:

- skip to the next field
- return to revise a previous field on the same record
- retain the value carried forward from the previous record
- retain the default value
- repeat the previous field (if it is repeatable)
- insert a copy of text stored elsewhere

All of these require only a few key strokes and tend to minimize typographical errors while speeding input. Throughout the process the DBMS performs at least some checking as to field length and content, so that letters are not allowed to be recorded in numeric fields, for example. Some go so far as to check authority files and reject unauthorized terms. Also, minimum and maximum value limits may be enforced in numeric fields.

UPDATING

Updating includes processes that change existing files. Sometimes it may be understood to cover input, sorting, indexing, and combining files, all topics that we have chosen to treat separately. This section discusses the processes of appending records, inserting records, local field changes, global changes, deletion of records, deletion of files, and—believe it or not—*un*deletion.

Appending adds records to the end of a file. If the file is kept in any kind of alphabetic or numeric order and records are not added exactly in order, then new records have to be put between old ones. This is known as *insertion*. Inserting records displaces those that follow—a new record 3, for example, makes the old record 3 into record 4, and old 4 into 5, and so on. Needless to say, this complicates the maintenance of permanent indexes.

All DBMSs support local field changes, the substitution of one value for another in a designated field of a certain record. For example, an object accessioned as a *stone* figurine may have been entered with *srone* in the Material field. A local change will replace this with *stone*. Later examination may determine that the object is really of ceramic material, requiring yet another change. Local changes are simple in fixed-length fields, but in a file of variable-length fields, substitution of a longer term for a shorter one may involve the system in considerable work—all behind the scenes, of course. If permanent indexes are maintained, local changes in a data file mean corresponding change to the index. In the example given, the changed record would be added to the list of stone object records and removed from the list of srone objects, and—it is to be hoped—*srone* could be deleted from the index altogether.

A global change applies not to one record but to *every* record with a designated value in a designated field. This facility can be very convenient when the museum changes its mind about preferred terminology, as for example, when a country has changed its name and all references to, say, *Tanganyika* change to *Tanzania*. Where large files are to be updated by sequential processing, such global changes may be done off-line, perhaps as an overnight run. Global change can be very dangerous, though. For example, a change from *Rembrandt* to *Van Rijn, Rembrandt H.* in the attribution field affects all occurrences of *Rembrandt* in names, producing such entries as *Peale, Van Rijn, Rembrandt H.; Soyer, Van Rijn, Rembrandt H.;* and even *Van Rijn, Van Rijn, Rembrandt H. H.* For this reason, many DBMSs do not support global change. Some, however, offer an alternative to such "blind" global changes. With such a system global changes are performed on-line. Only one change command is issued, but each successive instance is brought to the screen and executed only when the user confirms that the change should be made. Confirmation requires only a single keystroke per change.

Record deletion simply removes a designated record from a logical file. File deletion removes an entire file from the data base. Since many files are created for only temporary use, file deletion is a common procedure. However, because accidental deletion of a permanent file can be a serious loss, file deletion can often be done only by a user whose password confers special authority.

The misbegotten term *undeletion* refers to the ability to recover a record or a file that has been deleted. This is possible because deletion does not actually destroy anything. A deleted record is merely marked as deleted and remains in the physical file until it is sorted or otherwise reorganized. File deletion simply marks that file's references in the directory and/or table of contents and adds its space to a list of areas available for reuse. Until the space is actually used for something else, the file remains in storage. For a limited time, therefore, it is possible to undo a deletion. Not all DBMSs have this facility.

SORTING

Sorting is basic to all record keeping, manual or automatic, but, like many basic procedures, is not always simple as one might think. The objective is a series of records in alphabetic or numeric order by some aspect of their content.

DBMSs go about this in many different ways and offer the user a variety of options, all of which must be clearly understood if the result is to be useful.

The input to a sort—the set of records to be sorted—may be one file, a set of several files with identical or similar record formats, or a selection of records from one or more files. In the first case, the file may be sorted in place; individual records are simply moved up or down to produce the desired order. The file retains its original name and occupies the same space as before. Its length may be unchanged but, if some records are marked for deletion, the omitted records would result in a shortened physical file.

The result of a sort may also be a new file with a new name and physical location, leaving the original file(s) unchanged. This output uses about the same amount of space as the total input. A third possible output contains partial records. In other words, the sorting procedure may be instructed to transfer only certain fields of each input record to the new, sorted file. A fourth technique produces only a list of record numbers. Thus if five records containing the dates 1914, 1946, 1911, 1913, and 1899 were to be sorted by date, the output by record number might be 5,3,4,1,2. Using this list, the system can handle the original file as if it had been sorted even though no records were actually moved. The "signature" of this technique is a new file using very much less space than the original.

The DBMS may offer sorting options concerning selection, permutation, missing data, duplicate records, tie-breaking, and editing, as well as a choice of ascending or descending order. A *selection* option permits the user to specify that only certain records are to be included in the sorted file, e.g. only objects of US origin dated after 1899. *Permutation* applies to records with repeatable fields. For example, in a sort by former owner an object with three recorded former owners would be listed separately under each one. The case of *missing data* is illustrated by the object with no recorded former owner. Such objects can be omitted, brought to the top of the list, or relegated to the bottom. Some DBMSs offer the user a choice. Sometimes, especially when two or more files are sorted together, or when one file contains duplicate records as a result of past permutation, the same record may occur more than once.

Sorting can involve one in a great many unanticipated problems, some of which are so complex that there is no way a computer can always know the right answer, i.e., the answer that the human mind wants. One example is the difficulty in dealing with the "rank" of substrings within a character field. Familiar problems include

- leading articles ("The Playing Card Master")
- given names and titles ("Pieter Bruegel, the Elder" or "Baron James Ensor")

and any number of similar cases. What they all have in common is that only the human user can know that these strings should sort on the underlined letters as: "Bruegel Pieter Elder," "Ensor James," and "Playing Card Master." There are several methods by which the desired sort letter is indicated to the DBMS at time of input. Note also that internal punctuation has been removed in the sort strings.

VOCABULARY ANALYSIS

Unless vocabulary is strictly controlled from the outset, with every character field limited by an authority list, it is essential to analyze input early and often and root out inconsistencies before they become entrenched. Two common techniques for vocabulary analysis are the preparation of a KWIC (Key Word in Context) index for text fields and a frequency dictionary for other character fields. Both are explained and illustrated in the glossary. These functions may be included as features of the DBMS software. If they are not, the user must purchase or write separate applications programs for vocabulary analysis, and these programs must, of course, be compatible with the DBMS and the operating system in use.

COMBINING FILES

DBMSs are superior to simple storage and retrieval systems because of their ability to combine data from multiple files and so produce output that could not be obtained from any one set of data. Techniques for combining files include concatenation, merging, and joining (or relating). *Concatenation* links whole files end-to-end so that the last record of one is followed by the first record of the next. Each component file remains unchanged. *Merging* combines and sorts the records of component files so that the resulting file has the same ordering as the components. The result is the same as if two or more files were first concatenated and then resorted. Each record remains unchanged except that duplicates may be dropped and records previously marked as deleted may be completely eliminated. *Joining* is a more sophisticated process characteristic of relational systems. Individual fields from two differently organized relations are reassembled to form a new relation with a new record format. Only the fields remain unchanged and some new fields may actually appear.

Joining requires that the component relations have at least one domain in common, and this shared domain must be the primary key in at least one of the relations. Often it is a secondary key in the other relations. As an example of joining two relations, let us call the relation in which the common domain is a primary key *relation1,* since rejoining is possible only with a relational DBMS. The other will be *relation2*. What happens is that each separate occurrance of the common domain in relation2 becomes the nucleus of a new record in the resulting *relation3*. Therefore, relation3 will have the same number of rows as relation2. The rest of the relation3 format consists of fields

Figure 8
An Illustration of *Joining*

Artist (relation1)

	Name	Nationality	Birth	Death
row 1	Barlach, Ernst	German	1870	1938
row 2	Baumeister, Willi	German	1889	1955
row 3	Boccioni, Umberto	Italian	1882	1916

Works (relation2)

	Accno	Artist	Date	Medium	Title
row 1	656.39	Barlach, Ernst	1928	bronze	Singing Man
row 2	521.41	Barlach, Ernst	1927	bronze	Head
row 3	230.48	Boccioni, Umberto	1912	bronze	Development ...
row 4	231.48	Boccioni, Umberto	1913	bronze	Unique Forms...
row 5	507.51	Boccioni, Umberto	1910	oil	City Rises, The
row 6	142.53	Baumeister, Willi	1922	gouache	Composition
row 7	9.56	Baumeister, Willi	1955	oil	Aru 6

Works by age (relation3)

	Age	Artist	Accno	Title	Birth
row 1	28	Boccioni, Umberto	507.51	City Rises, The	1882
row 2	30	Boccioni, Umberto	230.48	Development ...	1882
row 3	31	Boccioni, Umberto	231.48	Unique Forms ...	1882
row 4	33	Baumeister, Willi	142.53	Composition	1889
row 5	57	Barlach, Ernst	521.41	Head	1870
row 6	58	Barlach, Ernst	656.39	Singing Man	1870
row 7	66	Baumeister, Willi	9.56	Aru 6	1889

Data taken from the first two files are combined to produce the third, relation3; data are from *Painting and Sculpture in the Museum of Modern Art*

associated with one value of the key in either relation1 or relation2, sometimes with the addition of computed fields (or columns).

In our example see (Figure 8), *Artist* is relation1, *Works* is relation2, and *Works by Age* relation3. The common domain is called Name in relation1 and

Figure 9
An Illustration of a Typical Report

PAGE HEADER

Nationality

Artist

Medium

Accno: Title

- -

ITEM LIST

German

Barlach, Ernst

bronze

656.39: Singing Man

Baumeister, Willi

gouache

142.53: Composition

oil

9.56: Aru 6

Italian

Boccioni, Umberto

bronze

230.48: Development of a Bottle in Space

231.48: Unique Forms of Continuity in Space

oil

507.51: City Rises, The

Data are from the files shown in Figure 8

Artist in relation2 and relation3. The Age column in relation3 is computed by subtracting Birth in relation1 from Date in relation2.

Relation3 might have included additional columns carried over or computed from the contents of the component files. The power of joining is clear, for relation3 expresses information that was not present, even in latent form, in either one of the component files.

REPORTS

A report is the visual representation of data, and a *report generator* is software that combines data with labels, punctuation, spacing, pagination, etc., to construct a document for human comprehension. Most DBMSs have a standard report format, which is little more than a column of field labels followed by field values. Others offer more than one version of the standard report format, while some include a report generation language so the user can design customized output with almost every variation that would be available to a layout artist, including a selection of charts and graphs generated automatically from stored data.

Perhaps the most useful report that does not involve computer graphics is the compressed or outline format (see Figure 9). This format presumes that data have been sorted, usually on a series of fields, and that sort fields carried forward from one record to the next are omitted.

PART III

COMPUTER STORAGE OF VISUAL IMAGES

8

Digital Images

Images have always played a role in documenting collections. Their value is the ability to describe what cannot be put into words. Small "recognition" photographs often appear on catalog cards. Drawings, diagrams, and high-resolution photographs are used routinely to record condition, document the progress of conservation, and even establish ownership when stolen objects are recovered. Photographic records also help to identify objects that have lost their markings. In some cases, images may actually substitute for objects under study, because many aspects of an object—such as pictorial content—are as clear in reproduction as in the original, sometimes clearer. Moreover, the images can be duplicated and mailed.

Sketches and maps, as well as photographs, are a valuable adjunct to a field collector's notes, which often become the foundation of object documentation. The techniques of image recording, storage, and display are also used for documents that include writing, with or without sketches or diagrams, especially documents with non-standard typography, exotic characters, or varying size and orientation of letters.

To digitize an image is to measure it in total detail: the distance from every point to every other point, the size of every angle, the curvature of every line, the area of every surface, and, in three dimensions, the volume of every shape—solid or void. All these quantities and all the ratios of one measurement to another can be extracted from a stored image automatically, without the curator's touching ruler or calipers or setting pencil to paper.

This automation of measurement is much more than a labor-saving fringe benefit, for it opens the way to quantitative and statistical investigations that have been foreign to the humanities. In recent decades, quantitative studies of style (though not of artistic merit) have come to play a major role in the study of literature and music, media that are more easily digitized than the plastic arts. With massive image bases, the same methods can be applied to the history of art.

IMAGE REPRESENTATION

There are two fundamentally different ways of reducing an image to a string of bits and bytes. The first is called the *schematic* or *outline image:* a series of points connected by straight lines. Curves are approximated by a series of short segments and any degree of approximation is possible. However, the shorter the segments, the more points must be recorded.

The requirements for recording an outline image are simple: two numbers to represent the *X* and *Y* coordinates of each point, plus something to show whether the adjacent line segment is to be drawn or skipped. This can be indicated, for example, by using negative numbers for a point that is not connected to its predecessor. In some systems the numbers may not represent point coordinates but lines, giving the length and direction of each successive segment. Either way, the image is recorded as a series of number pairs. If the outline image has tones of grey or color, there must be an additional series of numbers, one for each enclosed area, representing its coloration.

Three-dimensional outline images can be recorded in an analogous way. In three dimensions each point is represented by three numbers, for its *X,Y,* and *Z* coordinates, and the points are divided into groups, each group representing the corners of a plane polygonal surface.

The second approach divides the entire picture surface into an array of small squares, one or more numbers representing the coloration of each square. The squares are called picture elements or *pixels.* If the only values are black and white, a single bit expresses the color of each pixel and one million bits suffice to represent an image of one thousand by one thousand pixels. The more color variation, the more bits must be assigned to each pixel. Thus, four bits per pixel can represent a 16-level grey scale and twelve bits per pixel can represent any combination of 16 values, 16 hues, and 16 color intensities. Some systems go so far as to assign 48 bits to specify the color of every single pixel.

In three dimensions the image is composed of little cubes called volume elements or *voxels,* and each is described by one or more numbers corresponding to the visual characteristics recorded, e.g. color, texture, and transparency. A volume of picture space one thousand voxels per dimension would contain one billion (10^9) distinct voxels, an unmanageable body of data were it not for the image compression techniques discussed below.

INFORMATION CONTENT AND COMPRESSION

Every digitized image contains a certain amount of pictorial information, and this quantity can be given an exact numeric value, i.e., the minimum number of bits needed for a full description. This value has nothing to do with the apparent size of the image. In other words, the same image may be shrunken to the scale of a microdot or projected on the wall of a gymnasium without affecting its information content. The factors that determine information content are (1) resolution—how many distinct points (called *raster points*) the system can represent, (2) how many different color or other values can be assigned to each pixel or voxel, and (3) the complexity of the image—how many distinct edges, surfaces, etc., are described. Thus, the more curved lines are "smoothed" by using many short line segments, the greater the information content of the image.

It can be rather difficult to determine the exact information content of an image represented by pixels or voxels. However, for an outline image the problem is straightforward, as the following illustration will show. Imagine a flat, linear image of a triangle, without color, at a resolution of 1,000 by 1,000 raster points. It takes ten bits to express a number as large as 1,000. The triangle has three corner points, each specified by two coordinates, for a total of 60 bits. Three more bits are enough to specify that all three edges are to be drawn, not skipped. Hence, the information content is exactly 63 bits. If the enclosed area were to be colored with one of 64 possible hues, that would add six bits to the information content, for a total of 69.

Real systems often store and transmit more data than the true information content of the image. For example, since most computers process numbers in groups of 16 bits, a system might well store 32 bits per point instead of 20. To a certain extent, such waste of space can be justified by simplified processing when the image is displayed.

Now imagine this same, utterly simple image transmitted to a one-million-pixel bit-mapped screen for display. The terminal needs six bits to specify the color of each pixel—six million bits to display a 69-bit image!

Suppose that, instead of the solid colored triangle, the display were to represent an impressionist painting filling the entire screen. There would still be six bits per pixel, for a total of six million. The difference lies in information content. For the painting, almost every pixel would have to be described individually, not only for display but also in storage and in transmission. In other words the actual information content might be six million bits or close to that figure. That comes to three-fourths of a megabyte for one image; and a system that devoted not six but 48 bits to each pixel would use six full megabytes to store the image.

Image compression uses mathematical techniques to reduce the large amount of data needed to display an image to the often much smaller amount necessary for a full description, i.e., to the actual information content. For outline images this means storing only the coordinates of corner points and the colors of enclosed shapes, as described above.

Image compression applied to an array of pixels can be very complicated and may only approach the theoretical minimum. All such techniques rely upon the fact that many of the pixels comprising an image are often the same. Where this is true it may be more economical to describe each *type* of pixel only once and then to indicate where such pixels occur in the image. For example, the digitized image of a printed page may commence (in the upper left corner) with a one-inch top margin, and this may amount to a series of perhaps 100,000 consecutive white pixels. Obviously, it saves space to describe one white pixel and indicate that it is repeated 100,000 times. Spaces between lines would also contain thousands of consecutive white pixels, and even the width of a space between letters will have more than one or two. By taking advantage of such repetitions it is usually possible to reduce the bit content of a black and white page by 90% or more.

The main thing to remember about compression is that it depends upon the nature of the image. One with relatively little information content can be compressed greatly, stored in relatively little space, and transmitted in relatively short time, while another—such as a color reproduction of a painting—with enormous information content resists compression, requires a great deal of storage space, and takes a long time to transmit.

Finally, note that three-dimensional images tend to be highly compressible. Though such an image may include a billion voxels or more, the great majority of them represent either empty space or hidden space inside solid forms. Only the voxels that define the surface of a form need be described, stored, and transmitted; and this may amount to only about 0.1% of the whole picture space.

DIMENSIONS

The dimensions of an image correspond to the quantity of numbers needed to specify each point. In monaural sound recording there is only one dimension, time; and a series of single numbers represent audible vibrations. It does not follow, however, that sound recording is economical of space, for high fidelity requires the representation of several thousand vibrations per second. At this rate, a half-hour recording may consume 100 million bits or more—on the order of ten or twelve megabytes. Where high fidelity is not necessary, as in a telephone conversation, this can be reduced substantially; but very little compression is possible with sound. Binaural stereophonic recording represents two values for each point in time, doubling the information content of the sound image.

Flat pictorial images with their *X* and *Y* dimensions have already been discussed. The 1,000-by-1,000-pixel resolution assumed in our illustrations is rather high, though not uncommon. Such resolution is comparable to that of a good 35-mm photographic negative and better than a print from such a negative. Most display screens in use today are rectangular rather than square, with anywhere from 250 to about 750 pixels per dimension, for a resolution on the order of 250,000 points per screen.

A three-dimensional image is very much more than a view in perspective.

An image recorded in three dimensions can be rotated and viewed from any direction, including the top and bottom, and perspective can be adjusted to correspond to any desired focal length. With appropriate software, objects represented in three dimensions can also be viewed in any desired illumination. A three-dimensional image cannot, however, be transmitted directly to the screen, which has only two dimensions. Instead, a projection of the image onto the screen, as seen from a specified direction, at a specified distance, and sometimes in specified illumination, must be computed. What you see is only the projection. By projecting a series of views the image can be made to rotate on the screen; but, since each successive projection must be computed, the rate of motion is inversely proportional to the information content of the image. A complicated object may seem to turn very slowly indeed.

Images recorded in three dimensions can also be viewed stereoscopically. Two projections, from slightly different viewpoints, are computed and projected in rapid alternation. The viewer wears special goggles so that each eye receives its own image and not the other. There is a potential museum application for this feature in exhibition design.

Motion pictures may also be regarded as three-dimensional recordings, having two spatial dimensions and one of time. Whereas sound requires a temporal resolution of some thousands of instants per second, a moving visual image can be seen without flicker at only 30 frames per second. This means two or three gigabytes (billions of bytes) for a ten-minute sequence in color at about the resolution of cable television. The number is enormous, but within the capacity of a 5 1/4-inch optical disk.

9

Image Base Management

IMAGE CAPTURE

Although it is possible to generate a digital image by typing a series of numbers, the information content of a useful image makes this impractical. A variety of more efficient mechanisms has been introduced. Outline images are captured by tracing or sketching on a flat surface. One of the first devices for this purpose was similar to a drafting table. Sound sensors were mounted at the corners. The "pen" emitted a series of ultrasonic beeps and the time delay between each beep's emission and its reception determined the position of the pen. The coordinates of each beep were recorded as a point. The same concept is used to trace images directly onto a specialized CRT screen with a stylus or even a finger. In this case the screen is swept by horizontal and vertical light beams just above its surface and points are recorded where the instrument interrupts one beam in each direction. Outline images can also be drawn onto a CRT screen using an instrument called a light pen, which senses and records its own location every 1/30th of a second.

The method commonly used for tracing today involves a hand-held *mouse,* a small vehicle on wheels or ball bearings, used to trace or sketch on a pad adjacent to the terminal. Motions of the wheels over the surface are sensed and transmitted, via a wire *tail,* to a screen, where corresponding movements of a cursor trace an outline. Software functions available to modify the images include erasure, tilting, smoothing of outlines, changing the scale of a form, "painting" a shape with color, changing the location of a form on the screen, replicating a form, and combining forms to build a composite image.

Flat source documents such as pages and photographs, in black and white or color, are recorded by scanning. This can be done in two stages by using a television camera to capture the image in analog form and then a device called an analog-to-digital converter to create a bit string corresponding to the analog image.

The same kinds of images can also be digitized directly by a scanner that sweeps a beam of laser light across the image line by line from top to bottom, recording how much light is reflected at one point after another. Recording in color requires three light beams of different frequencies. Each point at which light is sampled becomes one pixel. Ordinarily there are from 75 to several hundred pixels per inch of the source document, horizontally and vertically. At a modest 200-dpi (dots—or pixels—per inch) recording density, an 8-½ by 11-inch page would produce 3,740,000 pixels. As we have seen, a pixel may require from one to 48 bits, depending upon how precisely color is measured. The time needed to scan a page varies with the type of equipment used and with recording density. It can range up to half an hour.

What is recorded is, of course, a full display image, not a compressed version. In the case of facsimile transmission, the bits may be sent directly to a printing device at a remote location where reproduction may be simultaneous with scanning. If the image is to be stored for local use or later transmission, it will usually be subjected to image compression as discussed in the preceding section.

Three-dimensional images cannot be recorded so directly. Instead, they must be computed on the basis of multiple two-dimensional views. A number of lenses may be trained upon an object, from different directions, as in stereometric photography, or one lens may circle about the object or the object may rotate before the lens. In any case, the full three-dimensional image, recording the back, sides, top, and sometimes the bottom, as well as the front view, is a mathematical synthesis. Like other images it can be compressed, enhanced, and processed in other ways. It can be stored and transmitted, but it cannot be displayed directly on screen or paper. Only projections of the image can be computed and displayed.

CAT-scan technology uses X-ray or other penetrating radiation to synthesize images of the interior structure of mummies and other objects that a museum may not want to dismantle. Such images may be either two-dimensional plane sections or full three-dimensional constructs. An image of internal structure has far more information content than a mere surface image and is, therefore, much less compressible.

We cannot leave image capture without some discussion of *holography,* which has been used as an intermediate step in digitizing three-dimensional images. A hologram is a piece of film like a photographic negative, but the resemblance ends there. A photograph records the incidence of light at the focal plane of a lens, so each point on the surface of the film corresponds to one point of the scene depicted. Therefore, one can see the image in the developed film. The holographic image is created quite differently. The incident light is not focused, so every part of the image is "in" every part of the film. Remove part of the film and you blur the whole image, but lose no part of it! The hologram records not *where* light strikes the film, but how rays of light—some reflected from the object and some not—interfere, reinforce, and cancel one another in their passage through the thickness of the film emulsion. The information is preserved inside the emulsion, where it cannot be examined. The film looks like a cloudy sky, bearing no resemblance to the image, which appears only when laser light, shown through the interference pattern, acquires the same characteristics as the light that made the pattern. Viewing such light creates the illusion of seeing the object, not on the hologram but in front of it or behind it. The effect is that, as the viewer moves around, the image remains stationary and one sees it from different angles. One cannot, however, rotate the image or see it from the back. In fact, it has no back, because only one side of the original object could be illuminated during the exposure.

The only way to compute a three-dimensional image from a hologram is to photograph the projected image from more than one angle, and the only way to complete the image is to use multiple holograms. The computational task is monumental.

IMAGE STORAGE

Historically, the main obstacle to the computerization of images has been their enormous information content, especially in comparison to text and numerical fields. Recent advances in large-scale storage, high-speed transmission, and image compression have just begun to open the way for routine, cost-effective use of digitized images with museum records.

Recently, several DBMSs have introduced image data fields that can be stored as an integral part of a logical record together with the text and numeric fields. Such a field takes about 12 kilobytes, roughly one thousand times as much as an average data field, yet only about 2% of the space needed for a high-resolution image. Obviously, these fields are intended for "recognition shots" comparable to the small photographs that are often pasted to museum object cards. In fact, they are promoted for use in personnel files, where 12-kilobyte resolution is more than adequate to recognize a human face. Records containing image fields are stored like other data records on magnetic tape or disk. However, one megabyte of storage, which might otherwise hold 2,500 or more personnel records, holds only about 70 if each contains a 12-kilobyte image. In other words, the space requirement is increased by a factor of more than 30.

As we have shown, a high-resolution image may have an information content ranging up to six megabytes or even more, while compressibility may range from zero to 90% or more. It follows that an image data base of museum objects requires about a half megabyte per object, give or take a factor of ten (roughly one digit to the left of the decimal point). Thus fewer than 100 images may fill a 40-megabyte hard disk that might otherwise hold 100,000 catalog records. It is clear that magnetic disk storage is far too expensive for a high-resolution image base.

Optical disk technology has made the image base economically viable. It began with the videodisc (the spelling of *disc* with a "c" is traditional in the entertainment industry where the videodisc got its start), introduced as a television recording device. Originally there were two videodisc technologies, the capacitance disc and the optical videodisc. The capacitance disc was less costly, but used a spiral track like that of a phonograph record. This made it almost useless for still images, which are difficult to locate on a spiral and even more difficult to hold because that requires repeated reading of the same segment as long as the image is viewed. The capacitance disc was in direct competition with the much cheaper videocassette and soon disappeared.

The optical videodisc uses a laser beam to read analog data from a track consisting of microscopic pits burned into the disc's reflective surface. The data track is not a spiral but a series of concentric rings, and its capacity is exactly 54,000 frames, the number that comprises a 30-minute show. It is easy to find an image by the location of its track and easy to hold it by staying with the same circular track instead of jumping to the next. The optical videodisc is essentially a publishing medium because (1) the original recording or "mastering" is very time-consuming and expensive, (2) once mastered, the data cannot be changed, and (3) the disc can be reproduced cheaply in volume by stamping, just as phonograph records are mass produced. For still images, the disc has the capacity for full color and the resolution of television, which in North America and Japan is 525 horizontal lines. Two extra data channels, intended for binaural sound, can be used for text, which may, optionally, be shown superimposed upon the image. Display requires a television screen. Museums have used the optical videodisc in two ways: to create a research image bank for internal use, and to publish illustrated catalogs of picture collections. Sales, unfortunately, are limited to the owners of videodisc players.

The standard 54,000-frame videodisc is 14 inches in diameter. When the same medium is used for music it is called a compact disc and the standard diameter is 5¼ inches. The recorded signal is analog, at present, even for digitally recorded sound; however, the medium can just as well carry a digital signal.

Because videodiscs and compact discs carry an analog signal, the data cannot be processed by a computer. However, images on an optical videodisc can be indexed by content and track location, and, if the *index* is computerized, they can be looked up and displayed under computer control. Such a system is exactly comparable to the automated retrieval, by mechanical means, of microfilmed images.

Optical disks (spelled with a "k") are equivalent to videodiscs. The only difference is that data are represented digitally, as on magnetic storage media. Whether they represent text, numbers, images, or sound, the digital bits and bytes are computer data subject to the same processing and transmission as if they were stored on magnetic disks. There are, however, two very important differences. First, the storage capacity of optical media is very much greater, measured in gigabytes (billions of bytes) rather than mere megabytes. Second, the data on the disk cannot be changed: it is a form of "read only memory," or *ROM*.

Standard sizes for optical disks are 14, 12, and 5¼ inches, the smallest format being called a CD-ROM (for "compact disc"-ROM). Because data on an optical disk are recorded in the same physical form as on a videodisc, the CD-ROM is also a publishing medium, and many data bases are now commercially available in this form. Indeed, some are issued periodically by subscription. Publications in this form may contain visual images and passages of recorded sound—even video footage—as well as text, but such mixed-media disks are useful only to those with the appropriate output devices.

Optical disks can be purchased blank and "written" under computer control. The computer outputs the same signal it would send to a magnetic device, but the data are burned into the surface of the optical disk. Once written they cannot be changed, only retrieved as often as wanted. This is called "write once-read many" or WORM technology. It is appropriate for very large bodies of data that are not to be updated, and digitized images fit this description. Data written in this way can, of course, be displayed, copied, and transmitted digitally.

The erasable optical disk, which will behave like a magnetic disk except for much greater capacity, is still in the laboratory as this is written. The most promising technology at this time uses a laser beam to write data, not by physically pitting a reflective surface but rather by changing magnetic properties of material beneath the surface. Because they have no physical relief, these disks cannot be stamped out for mass reproduction. Hence, erasable optical disks may not be a suitable medium for publication. They will be a very economical warehouse for massive, on-line data bases, however, which can be updated. The storage capacity of such disks will be adequate for image bases; however, they may prove less satisfactory for images than the current technology because no magnetic medium has the nearly indestructible permanence of the optical disk as we know it.

IMAGE TRANSMISSION

The problem of image transmission is comparable to that of storage—the sheer quantity of information: the result is low speed and high cost. The route from one device to another consists of various stages separated by modems and relay stations, and each stage uses an appropriate medium such as wire, coaxial cable, optical fiber, microwaves, or radio. The slowest link in the chain of communication limits transmission speed.

All electronic signals are carried by waves of one kind or another moving at a little less than the speed of light in a vacuum. This velocity, however, has almost nothing to do with the rate of transmission, which depends upon the reaction times of devices along the way and upon bandwidth, a measure of how many bits can travel in parallel. Bandwidth, in turn, depends upon frequency, which varies enormously. Waves of alternating current in a wire carry only about 2,400 to 10,000 bits per second, but radio waves can carry millions of bits per second and infra-red light in an optical fiber may carry 50 to 100 million.

The range of speeds, multiplied by the variability of information content, means that image transmission times vary from almost instantaneous to a matter of hours for a complex image on a low-speed line. Facsimile transmission of a black and white page usually takes a few minutes even though the *text* of a page, in ASCII code, would cover the distance in a fraction of a second.

Some systems bypass the problem of transmission entirely, using duplicate sets of stored images, one at each end of the line. These may be in the form of microfilm, with automated mechanical frame selection; or the images may reside on videodiscs, again with automatic selection and display. Then only the storage address of the image has to be transmitted. The drawback is obvious: an image can only be "sent" to a location where a copy is already on file.

Another common practice, briefly mentioned above, is facsimile transmission. The image, as an array of pixels, may be transmitted line by line as the original is being scanned, and it may be printed simultaneously at the receiving station. In this case no complete digital image may be generated, for parts are received and printed before other parts have been scanned.

Display of an image on a screen, as well as most image processing, differs from image printing in that the entire image, not compressed but fully expanded for display, must be available all at once. Therefore, a receiving station must either find room to hold what may amount to several megabytes of data or receive an entire image in the short time it takes to fill a display screen.

Real-time reception and display, as in the case of television broadcasting, requires considerable bandwidth, which explains why there are so few channels available in the portion of the electromagnetic spectrum allotted to television. Such wideband channels can also be carried by cable, which, however, is expensive and useful only for short distances. The rate of TV transmission is on the order of 10,000,000 bits per second for a resolution of 67,500 pixels per frame and 30 frames per second.

The lines ordinarily used for long distance communications are either telephone lines or data lines. The latter are faster and more expensive, but still too slow for real-time image transmission. An image received over such a line cannot be displayed or otherwise processed as a whole image until the last pixel has arrived. Optical fiber lines, coming into widespread use as this is written, offer much greater bandwidth and are adequate for real-time transmission of images, including those with resolution far beyond the television standard.

10

Image Processing

IMAGE MANIPULATION

Data representing an image can, of course, be processed in any logical way. We have already seen that an image, not too rich in detail, can be compressed to save space and transmission time and decompressed for display.

Images are also processed for enhancement. Perhaps the most familiar instance is the clarification of images transmitted from space. These weak signals pick up a great deal of static or "noise" along the way and arrive in a virtually unintelligible state. Enhancement algorithms use statistical methods to separate noise and distortion from the essential image. They are based upon assumptions about the smoothness of lines and surfaces, and, therefore, a small amount of detail may be removed along with the noise. Enhancement can add no information. It only subtracts what is presumed to be extraneous and meaningless.

Images can be altered in many elementary ways such as right-left reversal, slanting, vertical or horizontal expansion or compression (as with a bent mirror), and reduction to outline. Not all such processes have obvious applications. However, such things as reduction to outline, increasing contrast, and magnification can bring out detail that might not be visible otherwise.

Images can also be superimposed in interesting ways. For example, two versions of the same picture can be combined, with some adjustment to scale. Then if one image is "subtracted" from the other, what remains on the screen is a "map" of their differences.

Image magnification, or "zooming in," is done in several ways. The simplest is enlargement. This is applicable to a flat image, including computed projections of three-dimensional images. Another so-called zooming technique substitutes a partial image for the whole. This entails recording and storing not only the overall image but also a series of details. It is commonly used with maps, often in conjunction with a touch-sensitive screen. Thus, touching a display of the United States may invoke a display of the state that was touched; touching the state map may call up a county, and so on. This procedure requires an extensive data base of map images.

True zooming implies not only magnification but a change of apparent focal length. To understand this, imagine yourself before a mirror with one arm extended forward. As you advance toward the mirror the image of the extended arm grows faster than the rest of the body. At the same time parts of the background disappear behind the body and other parts that were out of the picture come into view at the edges. This is possible only with an image recorded and stored in three dimensions. Like the rotation of a three-dimensional form in space, zooming requires a series of recomputed projections and may proceed very slowly if the image is detailed.

CONTENT RECOGNITION AND OCR

Over the past four decades the power of computers has multiplied thousands of times over until today it is often measured in *megaflops* (millions of FLoating-point OPerations per Second). Yet certain classes of problems remain as intractable as they were 40 years ago, or nearly so. The case of image recognition is typical. The human brain, with much slower circuitry, analyzes the whole field of vision many times per second, effortlessly, even when the mind is focused elsewhere. A computer, using three TV "eyes," can barely direct a robot to pick up a cube and will never know the difference between a horse and a pig in a picture.

This is not to say that no machine can ever do it, but only that what we call a *computer* is not the right machine (see Abu-Mostfa and Psaltis 1987). Compare your view of this image with what a computer "sees":

```
111111111
111101111
111100111
111000111
110111011
100000001
100111001
001111100
111111111
```

You may quickly perceive the capital "A," but you do not immediately perceive the 81 bits of which it is composed. The computer, on the other hand, sees this as:

. . . 111111111 . . . 111101111 . . . 111100111 . . . 111000111 . . . ,
etc.,

where each series of dots stands for about 1,000 other assorted bits belonging to other characters on the same line as the "A." The bits are in main storage but processing takes place in the central processing unit (CPU), which can only examine a few at once. We shall not attempt to explain the tedious steps required to read a line of type.

In the eye and brain, storage and processing are combined. The first layer of the retina captures the image all at once, not one bit after another. Each cell of the second layer connects to many at the first and responds to patterns rather than to mere points. By the time the third layer is stimulated contrast has been intensified, lines and edges have been identified, and areas have been associated with movement in various directions. So the analysis proceeds, layer by layer, through the retina, optic nerve, and visual cortex, even as the first layer of the retina takes in the changing scene. Thus every part of the image is present momentarily in every successive array of cells, and every one of the millions of pathways a signal may travel through these layers serves as a data processor, working in parallel with all the other paths. By the time your authors say "Hi, Bob" and "Hi, Dave," their respective brains have discounted the effects of distance, orientation, lighting conditions, posture, movement, costume, and facial expression to match perception with personal concepts and names that are permanently stored in ways not yet understood. No general purpose computer can do that—any more than one of us could calculate the millionth digit of *pi*. (In 1985, however, Thinking Machines Corp. of Cambridge, Mass., introduced the *Connection Machine,* a radically new device incorporating over 65,000 processors and a wiring pattern more like that of the brain than a traditional computer. Image processing is one of the tasks for which such a machine is particularly suited and it is conceivable, though not yet proven, that its technological descendants may someday learn to "see" picture content (see Hillis 1987).

Significant progress is finally being achieved in one circumscribed aspect of pattern matching: *optical character recognition,* or OCR. It is now feasible to scan a written document and convert the contents to character representation (i.e., into ASCII or EBCDIC code). As the reader can well imagine, the computation is horrendous. It cannot be done at an acceptable rate without a long series of presumptions about the page to be read: it is right-side-up; the lines are straight, level, and evenly spaced; every character is surrounded by white and every line separated from the next by a uniform band of white; and the writing is in a known alphabet and type font, i.e., a library of stored shapes. Fortunately, the demands of OCR systems are satisfied by many typewritten museum documents, so input via OCR is often a viable option, bypassing the labor-intensive, error-prone process of keyboard transcription.

Conversion to character representation not only makes text available for ordinary text processing, editing, and indexing, but also saves space. An average text page can be represented in ASCII by about 50,000 bits, whereas its black and white image may require a million bits without compression, perhaps 100,000 in compressed form.

IMAGE OUTPUT

As we have seen, digitized images are of two main kinds: outline images, defined by corner points, and images represented by arrays of pixels. Display techniques also fall into two categories: hard copy and temporary screen displays. Hard copy is produced on paper and film, including microfilm and motion picture film. Outline images can be traced with pens, sometimes using a selection of colored inks. The paper may lie flat on a flatbed plotter, which is like a drafting table with a mechanism for moving the pens, or it may be rolled around a drum plotter, which traces lines by moving the pen back and forth along the top of the roller while rotation of the drum provides movement in the up-and-down dimension. These devices are intended primarily for engineering drawings, graphs, and maps. On film the lines are traced with beams of light.

Text printing devices, including the impact line printer, the dot-matrix printer, and all kinds of automatic typewriters, produce black and white images by using characters as pixels. For example, a ''W'' makes a much darker pixel than a period. Few printers can produce more than 132 pixels per line, but images of any size can be printed on a series of separate strips or pages to be assembled later. This technology may be useful for such simple images as bar graphs, logos, and ''banner'' headlines. It can also produce amazingly subtle pictorial effects, on the scale of a photo-mural, but there is little practical application for such output.

Laser printers print nothing but black and white pixels, at densities ranging from 75 to about 400 dots per inch. Instead of using characters to represent pixels, they use pixels to build images of characters. In principle, a laser printer can produce any character of any language, in any style or font, at any scale, in any orientation on the page. In practice, however, each system using a laser printer has a limited repertory of characters, symbols, fonts, and picture elements such as straight lines and arcs, because a small computer program is necessary to translate any given item from its byte representation into a *shape*. The laser printer has its own intelligence and its own main memory, where entire page images are composed and stored while the required number of copies are printed xerographically. The laser printer may also receive and reproduce an image already stored as an array of black and white pixels, though unfortunately most present day printers do not have enough storage to hold a full-page pictorial image.

The ink-jet printer differs from the laser printer mainly in the way the image is imprinted on the page. Instead of using xerography, the ink-jet printer literally throws a droplet of ink, guided by a magnetic field, to represent each pixel on the page. By using multiple nozzles with inks of different colors, some ink-jet printers can produce colored images, something that is not practical with xerography.

Finally, hard copy can be produced indirectly by photographing a screen display. This is particularly useful for motion pictures, where each frame may take a rather long time to compute, even though it is to be projected for only a fraction of a second. It may take days to shoot a 30-second, computer-generated television commercial.

Three kinds of screen are used for display, though the first hardly counts as computer technology. This is the automated microfilm reader. The image on the microfilm may have been generated under computer control and the film may have been selected and positioned automatically, but the medium is a piece of film and the device is mechanical. Display is essentially the same as slide projection.

The usual terminal screen is a television tube with a character generator. In fact, some economically priced computer systems have only the character generator, which the user must attach to an ordinary television set. The character generator is a digital-to-analog converter that receives character data and, for each character, generates the analog signal to produce a character image at the right place on the screen. A given character always has the same size and shape and is displayed at one of the regularly spaced character positions on one of the regular lines. Thus the screen resembles a typed page with every character in its proper row and column. No shape not built into the character generator can reach the screen, nor is there any possibility of variable spacing, horizontally or vertically.

Systems designed for graphics applications include "graphics characters," shapes such as rectangles and triangles of various sizes and colors that can be used to outline and fill areas on the screen, creating quite good charts and graphs or rather crude pictures and maps. Software can generate such displays rapidly enough to project the moving images essential to computer games.

Bit-mapped screens are typical of modern systems in the higher price range, especially the class of super-microcomputers known as *workstations,* which are becoming more and more prevalent. These displays are comparable to laser printers in that the screen is finely divided into pixels, and the repertory of shapes that can be projected depends not upon a hard-wired character generator but on software, so that, in principle, anything can be displayed. The display device contains a substantial amount of main memory—the bit map—in which one or more bits are permanently associated with each individual pixel. One bit per pixel is enough for a black and white display, while multiple bits support gradations of shade, hue, and saturation. Thus the contents of a bit map are exactly the same as the format in which a two-dimensional image is captured and stored uncompressed.

In use, the bit-mapped screen is often divided into separate, often overlapping areas called *windows.* Each window corresponds to a separate computer operation. For example, there is often a small clock image, with hands, associated with the timekeeping function that ordinarily goes on behind the scenes. Since the bit map changes only as instructed by the computer, a window containing data from a terminated program remains on the screen until erased by another program using the same area or until the screen is turned off. Thus an image or other data can be retrieved and kept on display as long as wanted while other parts of the screen display current processing.

11

Image Access

Image access takes in the whole range of techniques used to find, retrieve, and study visual records. These techniques make use of a complex web of relationships among three very different kinds of things: objects, images, and subjects (see Figure 10).

An object may be, for example, an organism, a tool, or a painting. An image is the visual analog of an object or subject, for example, a photograph of an organism, tool, or painting. A subject is anything you can name to which an object or image may be related. An example is Napoleon III, who may figure in a painting and also, therefore, in a photograph of it.

An image access system (computerized or not) is designed to answer any combination of the following six queries:

1. Given an object—find its image(s) and describe them
2. Given an object—list its subject(s)
3. Given an image—find the object(s) it represents
4. Given an image—list its subject(s)
5. Given a subject—find the object(s) that represent it
6. Given a subject—find its image(s) and describe them

This view of image access brings out several interesting—and troublesome—aspects. First, the givens are all singular, the search targets all potentially plural. Second, objects, images, and subjects are three totally different concepts, requiring different kinds of logical records. Third, as a corollary of the second

Figure 10
Image Access Queries

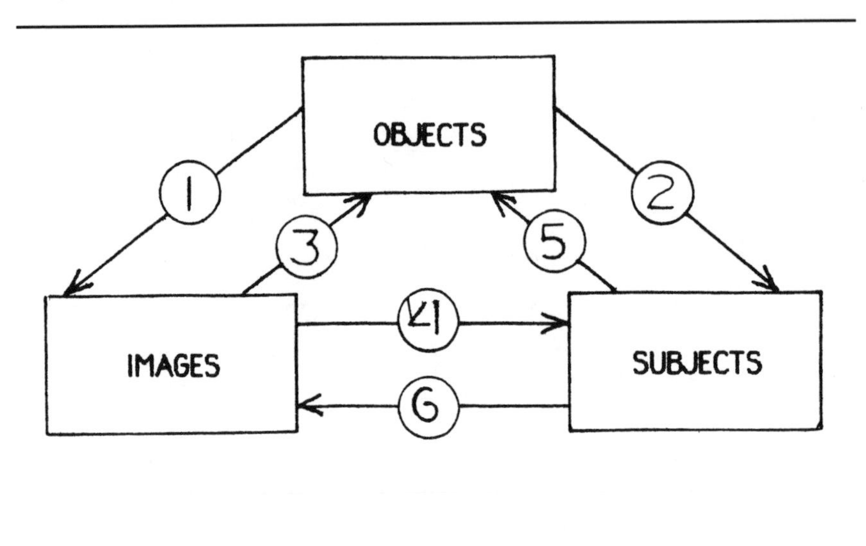

point, a portrait of Napoleon III is not one thing but two, an object *and* an image, requiring two different kinds of logical records.

Relationships between a single starting point or "given" and multiple targets are called one-to-many relationships. In diagrams such as Figure 10, they sometimes are represented by double-headed arrows from given to target. In manual documentation, one-to-many relationships are not especially troublesome. Likewise, a DBMS that supports repeatable fields or repeating groups closely emulates a manual system and accomodates one-to-many relationships, though at a cost in efficiency. However, there are only a few such DBMSs in use among museums. DBMSs that use fixed-format files, flat tables, and relational tables encounter problems with one-to-many relationships, hence also with image access. We shall see how these can be overcome using concepts already introduced under "List Processing and the Alternative" in Chapter 5 and data structures that may at first seem unnatural.

THE OBJECT RECORD AND IMAGE ACCESS

Museum object records were discussed and illustrated in Chapters 2 and 5, but with little reference to fields that support image access. Fields that point to images of the object (the first query in the preceding list) include references to photographic negatives by file number, references to published reproductions, references to conservation files that contain photographs, diagrams, and X-rays, and miscellaneous cross-references to other documentation that may contain

Figure 11
A Repeating Group of Pictorial Image Data

```
number:                 39.12.253
  .
  .
image_ref:              12,236
     function:              condition
     date:                  03/23/1952
     medium:                photo neg
     format:                5 x 7
     color:                 N
     detail:                UL head
     view:                  front
     published:             N
     location:              cons dept
     remarks:               raking light, after removal of
                            overpaint, before inpainting. Cf.
                            conservation report 52.7.
image_ref:              12,237
     function:              condition
        .
        .
  .
  .
etc.
```

images. Frequently these external references are collected in free text fields, where each reference may incorporate some indication of the type of image:

> photos: 12.34 8 × 10 B&W NEG; contact print of same in curator's files; 234.87 35 mm NEG in Reg Dept; see Cons Dept ref 73.4; color repro in Ptg. & Sculpt. frontispiece.

Such references are designed for human use. Automated retrieval of an image from a microfilm archive, videodisc, or optical disk requires a rigorously coded field referring to one image, not to a list of them:

> disc_track: 8765

Multiple images would then require multiple fields, a feature many DBMS do not support.

A system with repeating groups may include one or any number of rather complete image descriptions within the object record (see Figure 11).

The second basic image access query is, "given an object, list its subject(s)." Again, this is not difficult with manual records or a system with repeatable fields, but it is with most of the DBMSs in use. Six kinds of subject reference fields occur in object records: themes, generic subjects, proper subjects, literary references, Iconclass "signatures," and descriptive text. A single object may have any number of such references.

Themes (e.g. "Christmas") and generic subjects (e.g. "Christmas tree") are similar. Both appear as nouns and noun phrases; and in both cases the main difficulty is one of vocabulary control. For example, does "Christmas" stand alone or come under "holidays" or "festivities"? Assuming a well-organized thesaurus for this data category, the only problem left is that any object may require any number of subject references.

Proper subjects (e.g. "Santa Claus") refer to individual persons, places, structures, ships, horses, etc. The problem of vocabularly control is a little different than for generic subjects because the list of proper names is open-ended. An object may have any number of proper subjects.

A literary reference (e.g. "Moore, Clement Clarke (1779–1863), *A Visit from St. Nicholas*" or "Gen. 3:6–7") relates an object, usually but not necessarily a work of art, to a published text that is illustrated or interpreted. Multiple literary references are unusual but not impossible.

A so-called "signature" in the *Iconclass* system is a hierarchical code, composed of letters, numerals, and punctuation, that specifies the exact subject of a work of art, much as a Library of Congress number classifies a publication. However, a publication can have only one number, which also specifies its shelf location, whereas an image may have multiple Iconclass signatures (see Waal 1968).

Descriptive text is a verbal description of image content including interpretation, identification of proper subjects, and literary references, all in free text and often lengthy. For a motion picture the description is usually a synopsis of the story. Because of its bulk such a text may remain in a museum's manual files even when other parts of the records are automated. When automated, the descriptive text is subject to word indexing; and, in the absence of other subject fields, such an index provides the only reference to specific themes, generic and proper subjects, and literary references.

It is customary and almost logical to treat an image that is *on* an object as an integral part of the object, i.e., to include image fields within the object record. However, as noted above, image and object are, in fact, different entities, to be described in different terms. This subtle distinction becomes patent and troublesome in the common case of an object that bears a number of different images, for example a piece of furniture with decorative plaques. Such multiple images can be incorporated into the object record only if the system supports repeating groups.

THE IMAGE RECORD AND IMAGE ACCESS

An image record, roughly comparable to the repeating group illustrated in the preceding section (see Figure 11), covers three main topics and often a fourth: image description, description of the image as an object; reference to the immediate subject; and, for a representational object such as a work of art, references to subjects present in both the object and the image.

Under image description we include specification of spatial resolution (e.g. points per inch), color resolution, and the nature of the image (X-ray, raking light, infrared, schematic outline, etc.), all attributes with a bearing upon its purposes and use.

The image as an object may be a photograph (negative, positive, or transparency), hologram, drawing, sketch or etching, published reproduction, microfiche, videodisc track, or a digitized image on an optical disk or CD-ROM. Perhaps the most important aspect is some indication of where the image is, for manual or automatic retrieval.

In museum collection records the immediate subject of the image is usually a museum object, part of an object, or a group of objects (e.g. an installation shot). It is important to know whether the image represents the front, back, or side of the whole object or a detail, and whether it shows the object as it would be exhibited or, for example, stripped down in the course of conservation.

Finally, the image of an object or person, for example a portrait of Napoleon III, may represent that object or person, but a photograph of the back or, perhaps, of a damaged area, probably would not. The image record must answer the third basic query listed at the beginning of this chapter, "given an image, find the objects(s) it represents," and, where appropriate, the fourth query, "given an image, list its subject(s)."

Note that, even though most images in museum files represent only one object, the problem of the one-to-many relationship can and does arise. Where the image reproduces some or all of the object's subject matter, multiple subject references are normal.

SUBJECT RECORDS

In general, there is no such thing as a subject record *per se*. If the museum keeps (or has access to) biographical files, the persons listed may relate as proper subjects to objects and object images. The same may be true of a file of historic buildings, monuments, ships, show dogs, or race horses. Or a subject file may be nothing more than the inverted index to subject fields in the file of object records. What all "subject records" have in common is that they serve collection-based research by answering the fifth and sixth image access queries ("given a subject, find the object(s) that represent it"; and "given a subject, find its image(s) and describe them").

Obviously, the familiar problem of one-to-many relationships arises again. It haunts every phase of image access. On the face of it, this seems to imply that image access is possible only with manual files or DBMSs that support repeating groups. Fortunately this is not the case.

SUBJECT ACCESS AND FIXED FORMAT FILES

The majority of DBMSs in use today do not support either repeating groups or repeatable (multi-valued) fields. Therefore, they cannot accommodate indeterminate numbers of references from object to image or image to subject. This does not imply, however, that such systems cannot be used where image access is an issue. It does require a purposeful organization of files along lines unfamiliar to most users.

The answer is a system of intermediate files functioning rather like indexes, although they are not technically index files and may, in fact, be indexed. To illustrate, imagine a system including separate files of objects, images, and subjects, respectively. The subjects in this illustration are biographical entries. One of the objects is a painting representing the signing of the Declaration of Independence. The image inventory includes numerous photographs of this painting, many of them being details showing only one or two of the founding fathers. All files are in a fixed format, so the object record cannot include either a list of the photographs or a list of the persons portrayed. It is possible, however, to set up a fourth file relating objects to images, a fifth relating objects to subjects, and a sixth relating images to subjects. Each of these files would have a fixed format of just two fields:

File 4: Object-to-Image
object_no: 88.234
image_no: 34,567
etc.

File 5: Object-to-Subject
object_no: 88.234
prop_subj: Hancock, John
etc.

File 6: Image-to-Subject
image_no: 34,567
prop_subj: Hancock, John
etc.

This intermediate file approach entails removal of all multiple, external reference fields from the object, image, and subject files proper. In other words, the object record would have no field for theme, generic subject, proper subject, or Iconclass signature. Only the free text description and probably the literary reference could remain, because the former is not a repeatable field, while the need for more than one literary reference is extremely rare.

Does it follow that intermediate files (e.g. object-to-subject) should contain a series of subject fields such as this?

object_no:	88.234
theme:	American history
gen_subj:	group portrait
prop_subj:	Hancock, John

A little thought will show that this is not a sound idea, for the necessary permutations would set off an explosion of highly redundant data files. Therefore, record formats of intermediate files should be limited to a single pair of fields, as shown at the beginning of this section. Where the data includes several kinds of subject reference, each should have its own file.

Let us see how this mechanism responds to the six image access queries. In the fully developed system we have six files, or at least six classes of file: object file(s), image file(s), subject file(s), and the three classes of intermediate files illustrated above. The first step is always a query against the appropriate intermediate file(s).

Suppose we are looking for works of art that depict John Hancock. This is query type five, "given a subject, find the object(s) that represent it." The appropriate intermediate file is the one linking objects and subjects and the query is

FIND (prop_subj = 'Hancock, John')

Depending upon the type of software used, the response to this query is either a listing of object numbers or, better, a working file selected from the object-to-subject file, consisting of all records with "Hancock, John" in the prop_subj field. The user might name this temporary file "John."

This is a start, but a mere list of object numbers is hardly a satisfactory answer. Just what are these objects? That requires a query against the object file(s). In a system without the joining facility this would be a request to print or display all the object records listed by the first query.

Using a relational DBMS with the joining facility, it is very much simpler (see "Combining Files" in Chapter 7 and Figure 8). The file John and the object file(s) have a domain in common: the object_no field. Therefore they may be joined. This amounts to adding a full object description to each record in John, and the query is answered. If the file is long the user may go on to process it in other ways, such as sorting or selecting objects in good condition and available for exhibition; or another image access query may be initiated to locate reproduceable photographs of selected objects.

All image access queries may be answered in the same way: by an initial query against one of the intermediate files, followed by a query against the object, image, or subject file(s) or, if the system supports it, a joining of these files to the output of the first query.

PART IV

PUTTING IT ALL TOGETHER

12

Networks

A network is a system of connections. Basic networking terminology comes from a branch of mathematics called graph theory, which deals with patterns of "connectivity." These patterns consist of points called *nodes* joined by lines called *edges*. Elementary network forms (see Figure 12) include the star network, with one central node connected to several terminal nodes, the ring network, with each node connected to exactly two of the others, and the randomly connected network, which, unfortunately, is called a "network." Its characteristic is that at least some pairs of nodes are connected in more than one way; a road map is a typical example of this. More complex networks combine these elements to form trees (stars of stars), rings of rings, rings of stars, networks of rings and stars, and so on.

In a communication network—the only kind that concerns us—each edge represents a path for the flow of information and the direction of flow is shown by arrows pointing in one direction or both. For example, Figure 13 shows a star network representing the typical one-person office, viewed as an information network.

The same diagram represents a "stand-alone" computer system if the nodes are renamed Input, Computer, Storage, and Output. In the case of a small microcomputer, the whole network (keyboard, processor, disks, and screen) may share the same cabinet.

A large, randomly connected network may also be called a grid, analogous to grid of power lines. Often a smaller communication network is overlaid upon a grid of connections that is also used by others, as a beer truck's delivery

Figure 12
Examples of Different Types of Networks

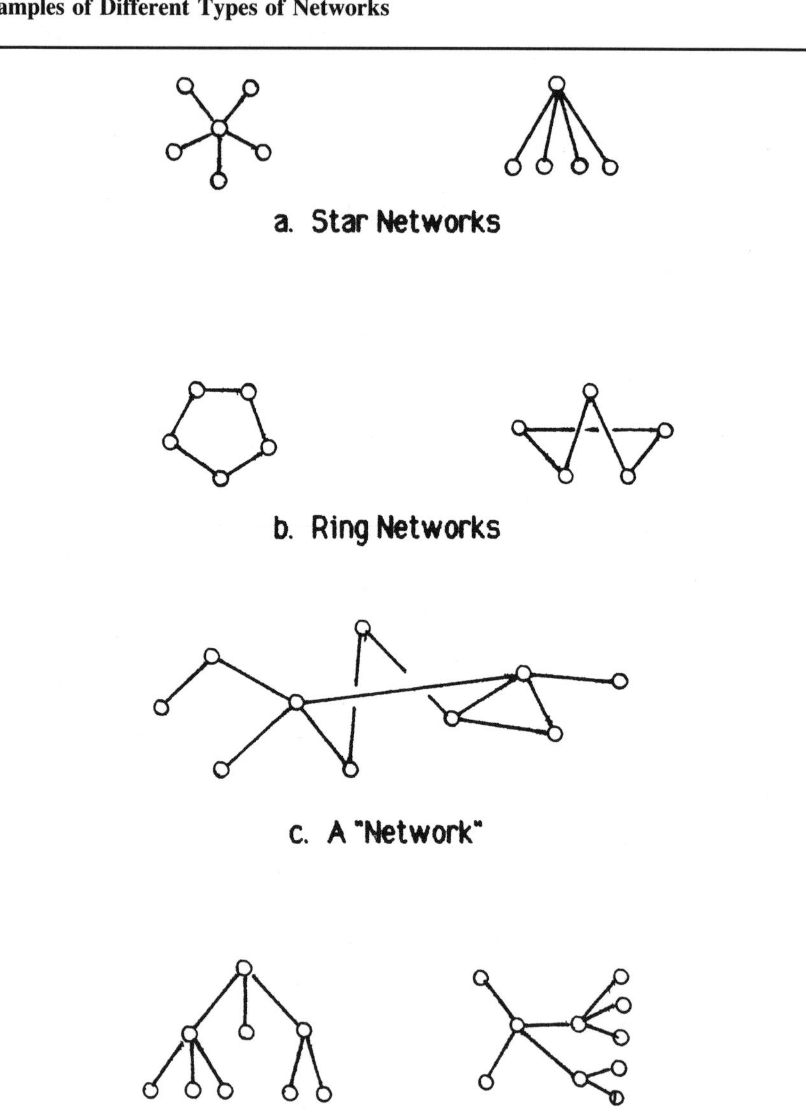

route (a ring network) makes use of the grid of public roads. Most computer networks that include more than one site utilize the grid of public telephone lines in the same way. So do informal networks of verbal communication among friends and colleagues.

Figure 13
A Network of Activities in a One-Person Office

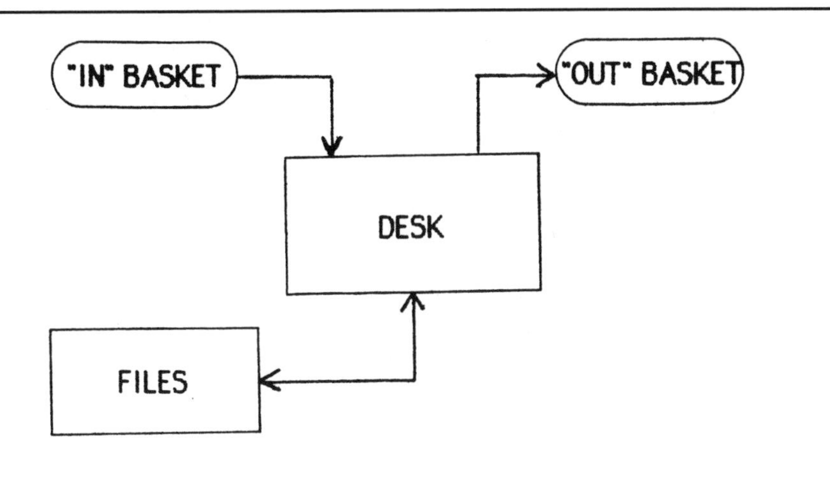

COMPUTER NETWORKS AND PROTOCOLS

A rudimentary computer installation with no external communications and only "dumb" terminals is a star network; and, like all star networks, it is also a *host* network. This means that one node, the computer, is dominant, either sending or receiving every message in the system and also controlling the timing and the format of every communication. If a second, independent computer is linked to the system, either locally or remotely, the result is a *peer* network, characterized by communication among equals. With equality come the risks of everyone speaking (or listening) at once, messages gone astray, varying rates of transmission, and even differently coded data, as when one sends in ASCII and the other tries to receive in EBCDIC. *Protocols,* embodied in shared network software, provide the "Robert's Rules" for all networks that include more than one processor. Usually such software divides each message into one or more standard-sized "packets," each including a coded address to one or more network nodes and a return address. A ring network is an example of a *broadcast* network in which all nodes receive all message packets, but ignore those addressed to others. In a *token* ring network one node sends while all the others receive, then passes the right to send (the "token") to its neighbor in the ring. The widely used Ethernet configuration from Xerox Corporation is a tree network in which any node may broadcast at any time. Simultaneous transmissions are "in collision" and must be rebroadcast until each has been clearly received and acknowledged. In other networks, messages must be repackaged and forwarded by nodes along the way. Most messages are echoed back to the sending node, which retransmits until a correct copy comes back, then sends its acknowledgement and waits for that to come back. Network protocols supervise this entire process, including many details beyond the scope of this book.

In microcomputer systems the networking software is often stored in ROM on a networking "board" plugged into one of the expansion slots. Note that each node must have its own copy of the protocol software, just as each student in a lecture hall must know the language of the lecture. Without it communication is impossible.

Today many computer installations, including relatively small systems used by museums, are designed as networks. They are not a single computer with peripheral devices but a community of linked machines, several of which possess "intelligence" in the form of a processing unit, main memory, and sometimes external storage. When all the machines are close together, as on a single campus, and communicate without long distance lines, the system is called a *local area network* or LAN. Otherwise they comprise a *wide area network* or WAN. The intelligent devices in such an installation may include (1) a central or host computer, (2) one or more other large computers, (3) a number of workstations—supermicro- or minicomputers capable of working alone—and (4) intelligent terminals, having some processing capability but unable to operate without the help of a computer or workstation. Most workstations have their own disk drives for external storage and some may control other peripherals such as printers. Additional workstations, called diskless, rely entirely upon storage shared by the entire network. In fact, it is characteristic of LANs and WANs that most storage and other peripheral devices are shared and used in turn by all the linked computers. When most or all of the network's external storage is pooled, the storage system is controlled by a dedicated workstation called a *file server,* which acts as the system's librarian. At the price of some added communications within the network, a file server greatly enhances the security and integrity of the data bases. If there is a central host computer, it usually controls the main body of external storage, often with the help of a *data base machine,* which is a powerful, special purpose computer that handles data storage, search, and retrieval operations for either a host computer or an entire network.

Data processing in a LAN or WAN environment may be carried out within a single computer or, when additional power is needed, transferred to a larger unit or divided into a series of steps or *tasks* parceled out to a number of machines. This division of labor may, in fact, be automated so that the user of a workstation may not know where the processing is actually done. This is called distributed processing and it involves movement of both data and software within the network.

In the same way, processing may involve data from files stored in several different locations, e.g. on disk drives associated with individual workstations. Related files under the control of separate computers constitute a *distributed data base.*

OPEN NETWORKS

Many small systems and nearly all large ones are connected to the worldwide public grid of telephone connections, private systems of higher-speed lines, or both. In a sense all such connected systems belong to one vast network. In another sense they do not, for there is no organization to the universal network. It serves rather as a grid, within which it is possible to create formal and informal open networks. The nodes are local installations such as individual computer systems, LANs, and WANs.

The most elementary example of an open network is one computer system calling another on the telephone. The ''network'' lasts as long as the call and the caller pays the phone bill. There is no administrative structure. At the next level of organization the called installation provides some regular service to which the caller subscribes. Typical services include ''bulletin boards,'' electronic mail forwarding, data base access, and actual computing power, i.e., time sharing. Although some such services may be offered gratis, as a kind of hobby, most must be paid for. This entails an administrative structure and some form of contractual arrangement between the server and the served. Ordinarily the subscriber pays a periodic fee to obtain an account number and access code, i.e., a password. Public line charges are the subscriber's too, as are costs of computer time, data storage, and royalties for access to data bases owned by others. The central system monitors subscribers' use of the network, computing costs automatically and distributing them to the proper account according to access code. Usually it even prints and mails invoices automatically.

A major example is Edunet, a subsidiary of EDUCOM, the Interuniversity Communications Council, Inc., a very large, international, nonprofit consortium of universities. All member institutions have remote access to powerful data processing resources offering every imaginable network service, including massive data bases and many program packages. A large subset of members that have major computing facilities provides the services. The Edunet administration in Princeton, New Jersey, coordinates these activities, arranging for the use of communications lines through commercial data carriers, collecting itemized charges from the universities and commercial vendors that provide services and sending each user a periodic consolidated invoice for all network use. Some years ago the Edunet administration informed the Museum Computer Network that museums might well be considered elegible for Edunet membership but, to our knowledge, no museum has applied. It is not known whether the offer is still open. Of course, museums associated with member universities have access to Edunet, insofar as their budgets can bear the costs.

The library profession accepted computing and then networking long before museums. This is not surprising in view of the relative uniformity of library training, cataloging, and procedures and the fact that most library objects are published—hence largely replaceable and interchangeable—while most museum objects are unique. Once one library has catalogued a publication it is unnecessary for another to do so, provided the second has electronic access to the first library's catalog. It is unnecessary, either, for a library to acquire everything published, provided it knows where publications are available for interlibrary loan. In sum, libraries have a great and obvious need to see one another's files. Many museums have indirect access to library computer networks through their own libraries or, in the case of university museums, through the university library.

The library profession has achieved a high level of standardization (at least in comparison to museums) not only in the cataloging of publications but also in the usage of personal names. This is not limited to authors' names but, by way of subject files, extends to all names, including biblical, mythical, and fictional personages who have been written about. This is the same population of names that inhabits museum records, so museums would do well to consult the library networks before deciding upon a standard form for any personal name.

MUSEUM NETWORKS

Museum information networks include national inventories of cultural properties, disciplinary networks formed by consortiums of museums with similar research interests, central catalogs of museums under a common administration (e.g. the U.S. National Park Service), groups of museums sharing software and standards (but not always data), and groups of museums sharing information about the principles and practices of computerization.

There are a number of national systems, some mainly on paper and other with substantial collections of information in being, but none approaching completion—whatever that might mean for a national inventory. National systems vary enormously in their approaches, objectives, and even technology, reflecting cultural, political, and financial conditions. Among the most advanced are those of France, Italy, Mexico, and Canada, the last being perhaps the most instructive example.

The Canadian project began in the mid-1970's as the National Inventory Programme (NIP), under the auspices of the National Museums of Canada. The original objective was a central catalog of museum objects, to be maintained in Ottawa. This was subdivided into a number of broadly disciplinary catalogs, ranging from art to zoology. The five national museums were expected to participate, while other Canadian museums had the option of joining or not, at no cost to themselves. Several hundred elected to do so, though not all could be accommodated at first. Recording standards were set by disciplinary committees and promulgated, along with technical training, by the central administrative staff. The network configuration was a simple star with all processing and data storage handled by a host mainframe at the central node. The other nodes had terminal devices for input and output.

As the data bases grew, a number of problems became increasingly clear. The object records included not only data of public interest such as artist and taxon, but also of purely local interest or even of a confidential nature, such as storage location and price paid. Many records were also embarrassingly inadequate in the eyes of present staff members. For these reasons there was objection to data access by the public or even by other museums. Moreover, museum personnel showed remarkably little interest in collections other than their own. Thus the central computer served primarily as a service bureau for about 150 introverted institutions.

Early in the present decade, NIP was reorganized as the Canadian Heritage Information Network (CHIN). The data bases were divided into public and proprietary parts, with the latter accessible only to the owner. Gradually, the "dumb" local terminals were replaced with small computer installations doing internal processing locally, but able to call upon the power of the central host.

There are probably more disciplinary networks than any one person can know, especially in the realm of systematics collections. MARDOC is a network of maritime museums in the Netherlands. It is also a network with a permanent organization and headquarters. Other special subject networks may be very informal, as when a number of scientists with common research interests exchange data files by "shipping" them as electronic mail, using the facilities of Edunet or Arpanet (a WAN built in the late 1960's by Bolt, Beranek and Newman, Inc., for the U.S. Department of Defense and connected to many research centers) as a communications grid. A museum's administration may sometimes be unaware of its participation in such nameless, *ad hoc* exchanges. Nevertheless it is a form of museum networking.

Groups of museums often share a unified administrative structure such as the U.S. National Park Service, Parks Canada, the U.S. Army Museums, and many others. These groups are similar in many ways to the national museum systems of other countries and sometimes just as extensive. The organizations named, and many others as well, have instituted their own intramural museum networks, often as part of an effort to develop a central inventory. Since the objective is usually a tightening of central control, these administrative networks tend to take on the centralized star shape of the original National Inventory Programme outlined above, except that small computer systems have generally taken the place of dumb input-output devices at terminal nodes.

Some organizations operate physical networks without information exchange. Typically there is a shared computing facility using a DBMS and other software adapted for processing museum records and a set of mutually accepted data recording standards, but communication takes place only between the host and each terminal node, with nothing (except perhaps electronic mail) relayed from one terminal to another. The network functions as a time sharing service bureau and standard-setting body, with a view to possible data sharing in the future. A remarkable example is the Museum Documentation Association (MDA) in Great Britain. An elaborate system of recording standards developed by national committees of subject specialists under MDA auspices is published and promulgated throughout Britain and other parts of the world. These standards are designed for both manual and automated information processing in order to facilitate user museums' gradual conversion. The MDA also supports a museum software package called GOS, which may be used locally or, optionally, by the MDA to input, store, and process data for museums without proprietary systems. A number of other networks of this general type have grown up where a major institution (e.g. the Detroit Institute of Arts) has acquired an in-house computer, developed standards and software, and made these facilities available to its neighbors.

In the mid-1980's the Getty Art History Information Program of the J. Paul Getty Trust sponsored its Museum Prototype Project. The objective was to build an experimental intermuseum data base merging information about eight museums' European paintings collections—much as the Museum Computer Network had experimentally merged data about 12 museums' drawings collections in 1968–69. The Getty project was more ambitious, however, focusing upon scholarly research information rather than raw catalog data. A considerable body of information was collected in an orderly format and, at present writing, is still being analyzed and studied. The project itself has been terminated for reasons that are now familiar: the program was interested in art historical research, in line with its mandate, while each individual museum was concerned first of all with collection management. Some museums continue to use hardware, software, and experience gained from the project to control their general collections, not limited to European paintings.

The Museum Computer Network (MCN) is an example of museums joining together to share information about standards, principles, technology, and experience without actually establishing a physical communications network. Such was not the original intention. The MCN was formed in 1967 with the notion of setting up a local communications network of art museums in New York City. The host computer was to have been an IBM model 360/40 at the now defunct Institute for Computer Research in the Humanities of New York University, while intelligent nodes throughout the boroughs would have used the long forgotten 360/20 computer. The physical network was never funded. A software system called GRIPHOS, developed primarily for the United Nations Library, was adapted for museum use and distributed by MCN to a handful of institutions that used it for their own purposes from about 1969 until 1978, by which time such batch-processing software had become obsolete. No electronic data exchange ever took place. MCN today is an international organization of institutions and individuals, jointly and severally addressing the matters that are the subject of this book. Another international group discussing the same problems is the International Council of Museums Documentation Committee (CIDOC). Many of the same people are active in CIDOC, MCN, and MDA.

USES OF MUSEUM NETWORKS

The technical feasibility of networking and union catalogs was apparent 25 years ago and had been demonstrated at least by 1968, but in the quarter-century since then few museums have shown an interest in either contributing to such resources or making use of them. Economic factors have led to frequent sharing of facilities, software, and recording standards, but not to the exchange of even the most public information, the merging of files, or the development of public union catalogs.

It may be instructive to examine exceptions to the rule. They are seen, firstly, among systems of museums under unified administration and, secondly, among individuals and research departments (rather than whole institutions) with a common interest in narrowly focused research. In the former case the impetus for networking comes from the central administration in the name of economy, national pride, inventory control, and centralization of authority, with a formal bow in the direction of advancing knowledge. In the latter case, networking is motivated by scholars' pursuit of immediate research goals rather than creation of a permanent information resource.

We have seen that the economy of shared cataloging does not have the same allure for museums as for libraries. There are important exceptions, however, for not all museum objects are unique. Handmade prints and manufactured objects, often with identifying patent numbers and trademarks, are represented in multiple collections, entailing duplicate research efforts. Personal biographies and the histories of corporate and cultural entities are independently researched at great expense for many different museums' collection documentation. The benefits of pooling such information resources are obvious, but the administrative framework has yet to appear.

There is one remarkable success story, not at the center of the museum community, but at its periphery. This is the International Species Inventory System (ISIS—not to be confused with two software systems with the same name) sponsored by the American Association of Zoological Parks and Aquariums and headquartered at the Minnesota Zoological Gardens in Minneapolis. Since the early 1970's, this data base of animal populations in captivity and individual histories and pedigrees has become an indispensable tool for the preservation of specific "gene pools" and, hence, of some animal species. The zoos of the world can no longer get along without it and that is the key to lasting success.

The single most essential ingredient in the success or failure of efforts in museum documentation, especially where automation is a factor, is dedicated, long-term administrative backing. Loss of this backing almost invariably spells doom or, at best, a period of stagnation. Unfortunately, its loss is all but inevitable in the succession of institutional and governmental administrations. The problem is acute for the single institution, but very much more so in cooperative ventures such as networks that demand the permanent support of many independent organizations.

What then can assure permanent, enthusiastic support? Apparently only necessity, as in the case of ISIS, or near necessity, as in the case of library networks. Another motive, not yet tried, is profit, as suggested at the end of this chapter.

The ideal museum network would, of course, be a public union index of museum holdings, organized not according to the accident of current possession but rather by significant relationships of one object to another. Such universal compendia are not practical in the form of traditional publication on paper, for the market could never justify the expense, nor could an index be compiled, even for a narrow class of objects, before becoming grossly outdated. Only an on-line data base drawn continuously from the changing files of the world's museums would serve the purpose.

The fundamental difficulty with projects of this kind is that the principal benefit is to the scholarly community and the public at large, while the principle cost, under present arrangements, would accrue to individual museums and multi-museum organizations, which can barely afford their present services to a local public. In the future one may imagine profitable, value-added networks paying for access to collection records, merging and indexing documentation, and marketing a data base service beyond the dreams of any present museum network.

13

A System for Your Museum

In this chapter, the principles discussed previously will be brought to life by description of an actual case study—the catalog created for the museum at the Indian Pueblo Cultural Center (IPCC) in Albuquerque, New Mexico. The records of each artifact in the IPCC Museum were typed into computer memory using a relatively inexpensive personal computer and a data base management system that is available to anyone for nominal cost. This is the story of that experience. It does not illustrate everything discussed in the preceding chapters—that would be impossible—but, hopefully, it will provide sufficient information for an intelligent reader, unskilled in the area of computer science, to know how to proceed with the development of a museum system, with clues to the problems that can be expected, their resolution, how much time will be involved, and approximate costs.

The All Indian Pueblo Council is an organization of the 19 Pueblo Indian tribes located in the state of New Mexico. Headquarters for the council is the IPCC, a D-shaped, two-story building constructed in the form of an Indian pueblo surrounding a central plaza. In addition to the council meeting chambers and offices, the building houses other Indian organizations, a Pueblo Indian restaurant, a gift shop, an auditorium, and a small but very fine museum that is dedicated to displaying the distinctive characteristics and artifacts of the various Pueblo tribes. The museum is directed by a paid employee of the council, but most of the work is done by a dedicated group of volunteers who are members of an organization called the Friends of the Indian Pueblo Cultural Center.

Figure 14
Format of 3″ x 5″ Card Used to Record the Physical Inventory at the Indian Pueblo Cultural Center

Accession Number

Object Name

Pueblo (or other provenience)

Location (within the museum)

Value

Artist/Craftsman

THE ARTIFACT CATALOG PROJECT

About two years ago, one of us spent a few hours with the museum director and two or three of the volunteers who had become concerned about the number of artifacts for which the museum was responsible and the lack of control that anyone had over these objects. A decision was made to take a complete physical inventory of the collection, recording each artifact on a 3″ × 5″ card, with the format and data categories shown in Figure 14.

As the information was recorded, Accession Number was actually a four-part code. The first part of the code is a letter indicating what would normally be called "method of acquisition" ("L," "D," or "P" for objects acquired on long-term loan, as donations, or by purchase). The second, third, and fourth sections of the Accession Number are the standard form recommended by the American Association of Museums for "year of acquisition," "acquisition number," and "artifact number" (within each acquisition). Normally, "method of acquisition" would be considered as a separate data category, not connected to Accession Number. However, in this case, numbers originally assigned to individual artifacts included the initial letter "L," "D," or "P." This meant that the numbers alone are in some cases duplicated if the letter part of the code is not also included as part of the Accession Number.

The naming of objects appears to be simple. As will be seen, though, what starts out as a functional name designation of only one or two words is often expanded to include materials out of which the objects were created, condition, shapes, titles (of works of art), and subject matter (works of art and photographs)—all categories of information that should be separated from the Object

Name category. Almost all beginners tend to expand and confuse the structure of data in key fields in this manner rather than add new fields to control other categories separately or, at the least, to allow simple, alphabetic control of Object Names without multiple, adjectival subdivisions. Data categories dealing with different kinds of observations about objects should never be intermixed. To do so greatly diminishes the value of the file for sorting, selecting, controlling, and listing by computer (for a more detailed discussion of this problem see the first chapter of Chenhall 1978).

Artifacts that were known to have originated in one of the pueblos presented no particular problem, but where the provenience was somewhere else—say a prehistoric Indian site—the initial recording of the data was somewhat less than consistent. Sometimes it was a site name (e.g. Casas Grandes), sometimes a pottery type (e.g. Mesa Verde black-on-white), and at times something else again. We will see how convenient it is to use special listings of the computer file to pinpoint problems such as this and identify the records that need to be modified for the sake of consistency.

In the initial inventory, Location was recorded with one of only two letter codes: "M" to indicate that the artifact was on display in the museum, or "V" to show that it was in the vault, i.e., in storage. This information, of course, is so general as to not be particularly useful in the management of a museum inventory. The concluding section on "Plans for the Future" discusses how the file can and will be modified to make the location information more specific. The remaining data categories in the original inventory—Value and Artist/Craftsman—do not require discussion at this point.

CREATING A COMPUTER FILE

In order to create a computer file—a necessary first step to the accomplishment of ultimate objectives such as those described in the remaining sections of this chapter—four or five distinct and sometimes difficult things must be done:

1. An appropriate computer must be selected and installed, and someone in the organization must take the time to learn how to operate it.
2. An appropriate DBMS must be acquired, and someone in the organization (presumably the same individual or individuals who have learned how to operate the computer) must study the system carefully to be sure that it will accomplish the needs of the museum better than any other DBMS available for the computer that has been selected.
3. An *information system* (as distinct from a DBMS) that will answer the known, identifiable needs of the institution must be thought through and structured so as to provide, within the DBMS, the files and fields that are necessary to produce the permanent records. We are fortunate today in that most DBMSs allow the addition of fields that may have been initially for-

gotten. However, it is always best if the files and fields most probably necessary are provided before the initial data is entered.

4. Finally, one or more persons must transcribe or type the data on each of the artifacts in the collection from the manual (or typed) records into the DBMS for storage as machine-readable records.

Taken together, these initial activities are the most important determinants of whether you end up, not with a toy, but with an information system that processes essential museum information in the most efficient manner possible. The number of records involved and the complexity of the information needs should be thoroughly researched before any decision is made on any of the four points. We will provide as much information as possible so the readers can extrapolate from the IPCC experience to their own needs, but we do not recommend that anyone should copy what is described here. It is essential that each institution create the information system that it needs.

The computer used by the IPCC is an Apple Macintosh with one megabyte of internal storage capacity or RAM—this is the model popularly called the "Mac Plus." The only auxiliary hardware items are a second, 800-kilobyte double-sided disk drive, and the standard Imagewriter printer. The reasons this computer was selected were, very simply, the fact that it was available to us, and the fact that we believed the one megabyte of memory, together with the particular characteristics of the DBMS selected, would enable us to easily handle the size and complexity of the records we would be recording. We wish to emphasize that we have no association with Apple Computer, Inc., that we paid full retail price for the computer, and that other brands of computers used in conjunction with other DBMSs would undoubtedly accomplish the same result. We are simply describing our particular experience as an example of what is possible with the computers on the market today. Apple and Imagewriter are trademarks of Apple Computer, Inc., and Macintosh is a trademark licensed to Apple Computer, Inc.

SELECTING A DBMS

The problem of selecting the best available DBMS for a particular application is one of the most difficult and frustrating activities that museum personnel will face. Systems and versions come and go like mushrooms. Quoting from a letter one of the present authors sent to the other recently: "As for what any given system can do, we sail on a sea of lies. One cannot trust salesmen. One cannot trust promotional literature. One cannot even trust operating manuals—even if there were several full-time people reviewing them. Least of all can one trust museum users, who must always report success or have their budgets gutted. . . . Not that all are bad. There's just no way of knowing."

So, we include here as Figure 15 a list of DBMSs that have been used successfully by museums. Several of these program packages were developed

Figure 15
Some of the Data Base Management Systems Used by Museums

Acronym	Developer and/or User
ARTIS	Art Museum Association of America
DARIS	Detroit Institute of Arts
Dbase III	Ashton-Tate
FOCUS	Information Builders, Inc./used by Yale Center for British Art
GOS	Museum Documentation Association
INFORMIX	Relational Database Systems, Inc./use by Princeton University Art Museum
InfoStar	MicroPro/used by Emory Univ. Museum of Art
INGRES	Relational Technology, Inc/used by Peabody Museum, Harvard
INQUIRE	Infodata Systems, Inc./used at NY State Museum
MINISIS	Systemhouse, Ltd./used by Historic New Orleans Collection
ORACLE	Oracle Corp./used by Art Index of Denmark
PARIS	Control Data Corporation/used by Canadian Heritage Information Network
REVELATION	Cosmos, Inc/used by Mystic Seaport
SELGEM	Smithsonian Institution
SPIRES	IBM Corporation
STAIRS	IBM Corporation
ZIM	Zanthe Information, Inc/used by Ontario Museum Computer Network, Waterloo

by or specifically for the minicomputers or mainframe computers at the institutions where they are used. Others are generic systems created for a variety of users and for multiple computers.

The Clearinghouse Project at the Metropolitan Museum of Art (see under "Information Resources" at the beginning of the References section) was created, in part, as a resource regarding currently active museum computer projects. The serious reader would certainly want to make use of this. Museum Services International (also listed in the "Information Resources") is a private organization that evaluates the usefulness of particular DBMSs for museums. In addition, the authors have recently learned of a new publication entitled *How to Choose a Data Base* (Small 1987). Neither of us have had an opportunity to review this volume, so we cannot recommend it. For those who may be interested, it is available from the American Philological Association, Department of Classics, Fordham University, Bronx, NY 10458–5154, $4.50 post paid.

The DBMS used to create the IPCC catalog is called OverVUE. It was designed for the Macintosh computer by ProVUE Development Corporation, although it is also available for IBM and, perhaps, other brands of computers as well. Both OverVUE and ProVUE, of course, are registered trademarks of the corporation, with all rights reserved.

The reasons for selecting OverVUE were several. Each data sheet (this is the word that OverVUE uses for a file template) may contain up to 64 columns or fields of information. The number of bytes in each field must be specified before one begins recording, but it can be changed at any time. Most importantly, the field size is only an approximation of what one will want for printing purposes—you can always enter up to 62 characters in any field, no matter what is specified on the data sheet. In other words, within limits, OverVUE provides for variable-length data fields instead of wasting storage space by specifying fixed-length fields to accommodate the maximum probable length.

Another characteristic of this system that we consider important is the fact that sorting of data is done entirely within the RAM. In some systems, sorting can be extremely slow as data is transferred to and from the disk drives for processing. With OverVUE, records that are selected for processing (or the entire file) are moved into RAM for processing so that processes such as sorting are accomplished at true electronic speed. From the perspective of human response times, this means that these activities are virtually instantaneous. However, the other side of the coin is that such a process entails keeping the records being sorted in main storage (RAM); therefore, the total size of the file(s) to be sorted is limited by the availability of working space. This is not a problem when sorting is done on either hard disks or external disks. In the present situation, the files are small enough (and will continue to be so into the forseeable future) that file size was not a limitation.

OverVUE provides a constant tabulation of statistical data regarding the file that is in RAM. These statistics can be displayed at any time in the format that is reproduced (with actual IPCC numbers) as Figure 16.

Figure 16
Statistical Data Maintained as Files are Created with the OverVUE DBMS

Selected/Total Records :	1972/1972
Summary Records :	0
Average/Total Width :	44.8/94
Column Avg/Max Width :	0.9/1
Number of Columns :	9
Percent in Use :	10.0%
Characters Free/In Use :	789651/88177

These statistics show that the IPCC file contains 1,972 records (we will see the format of these records in the next section), that all 1,972 records have been "selected" and transferred into RAM, and that these records consume 88,177 bytes of storage or approximately 10% of the net space available in the one-megabyte RAM. For many purposes we will select only certain records or columns of information to be transferred into RAM from the disk storage, in which case the number of characters in use and the percentage of the available capacity will often be considerably less.

Summary Records are primarily for use in business applications and will not concern us. Average/Total Width means the average total record length (in bytes) and the total display length (in this book we refer to record "length;" in the OverVUE system record "width" is the same thing). Actual records, of course, may contain more or less characters than this value. Column Avg/Max Width is the length, average and maximum, of the particular column (field) that happens to be highlighted when the statistics are called up. In our example, this is clearly a one-character field, probably Method of Acquisition.

ENTERING DATA

Undoubedly, the slowest and most tedious activity in the creation of computerized file representing a museum collection is the record-by-record typing of the data into the computer. Certain characteristics of some DBMSs make this much easier than it is with other systems, but it is always the hardest part of any computerization project. In the project we are describing, approximately 30 hours were required to enter 1,972 records. Because of the tedious nature of the work, it was extended over a period of five weeks, with typing taking place no more than two or three days per week, one or two hours per sitting. If a collection consists of 20,000 or 200,000 objects, if data on each object are considerably more complex, or if the manual records are poorly organized so that data entry involves a good deal of interpretation as well, the time allotted for this activity must be extended accordingly. In addition, one must be constantly aware of the fact that accuracy deteriorates rapidly with increasing boredom. In summary, it is impossible to overemphasize the importance of careful planning of the data entry activity.

PRINTOUT OF THE FILE

From the statistical information displayed in Figure 16 we learned that the file we created contained nine columns, corresponding to the six categories of information on the 3″ × 5″ cards (Figure 14), with Accession Number divided into four parts. Once a significant number of objects have been recorded, the usual procedure is to print out whatever has been typed, in the same sequence and format that the information was recorded. In this example, we selected a small portion of the 1,972 records for display as Figure 17. In order to get the record onto an 8½″ × 11″ page it was necessary to print this in a semi-condensed format, which gives us a slightly smaller print than either pica or elite type.

You will note that some of the data appear to be truncated for lack of space. Some of the object names and some of the pueblo names, though understandable, are obviously not complete. This is because we set the lengths of the fields on the printing template to allow us to print out the whole record in semi-condensed format. Remember that OverVUE allows a variable-length field up to 62 characters on any entry in a particular field. The truncated information, thus, is not lost; it is just not visible when we print the entire record. As we will see, we can select a limited number of fields and print out even the maximum number of characters we have recorded in any one field.

SORTING AND PRINTING BY OBJECT NAME

The IPCC computer file we have described, recorded on magnetic disk and printed, in part, in Figure 17, should be thought of as *the* museum catalog, containing what we have called primitive data (see Figure 5). Changes will be made, of course. Especially at the beginning, there will be many kinds of corrections to be made and new records will be added to the file. Later on, there will be regular changes to keep track of the changing locations of artifacts. In time, some of the records will have to be deleted as objects are decessioned. Every time this file is modified in any way, a backup copy of the entire file should be made. The backup disk(s) should be stored in some location other than where the active disks are kept; and in every way possible, all museum employees must be trained to protect these disks, both from unintentional changes and accidental destruction. Any new files prepared from this master museum catalog will be what we described in Chapter 5 as working files.

With the IPCC catalog, the first working file that was prepared was an al-

Figure 17
A Portion of the IPCC Catalog in the Sequence Recorded

L/D/P	Yr.	Acc.No	Obj.	Object Name	Pueblo	Loc	Value	Artist
P	85	1	97	Pitcher	Santo Domi	V	35.00	
P	85	1	98	Tray, Bird	Santo Domi	V	25.00	
P	85	1	99	Vase, Polychrome	Acoma		35.00	
P	85	1	100	Bowl, With Handle	Santo Domi	V	35.00	
P	85	1	101	Bowl, Black Carved	Santa Clar	V	45.00	
P	85	1	102	Tumbler	Maricopa	V	35.00	
P	85	1	103	Pitcher, Red & Cream	Santo Domi	V	9.00	
P	85	1	104	Bowl, Incised	San Juan	V	25.00	
P	85	1	105	Wedding Vase, Red & White	Jemez	M	12.00	
P	85	1	106	Basket	Santo Domi	V	25.00	
P	85	1	107	Vase, Polychrome	Santo Domi	V	38.00	
P	85	1	108	Plate, Black & White	Cochiti	M	18.00	
P	85	1	109	Bowl	Hopi	V	25.00	
P	85	1	110	Bowl, Bird	Acoma	V	8.00	
P	85	1	111	Prayer Basket	Santo Domi	V	75.00	
P	85	1	112	Vase	Sioux	V	3.00	
P	85	1	113	Bowl, Polychrome	Isleta	M	20.00	
P	85	1	114	Jar, Black	Santa Clar	V	85.00	
D	85	2	1	Photos, "Firing"	San Ildefo			
P	85	2	1a	Moccasin	Laguna	M	50.00	Purley, Leonard
P	85	2	1b	Moccasin	Laguna	M		
D	85	2	2	Photo	San Ildefo	V		
D	85	2	4	Photos, "Popovi Da"	San Ildefo	V		
D	85	2	5	Photo, "Popovi Da"	San Ildefo	V		
D	85	2	6	Photo, "Popovi Da"	San Ildefo	V		
D	85	2	7	Photos, "Maria, etc."	San Ildefo	V		
D	85	3	1	Lithograph	Santo Domi	V	400.00	Lovato, Charles
P	85	3	1	Bowl	San Ildefo	M	30.00	Sandoval,
D	85	3	2	Lithograph, Artist's Proo		V	800.00	Redstar, Kevin
P	85	3	2	Bowl	San Ildefo	M	30.00	
D	85	3	3	Serigraph		V	800.00	Redstar, Kevin
D	85	3	4	Serigraph, Proof #5		V	800.00	Redstar, Kevin
D	85	3	5	Serigraph, Proof #6		V	800.00	Redstar, Kevin
D	85	3	6	Serigraph		V	800.00	Redstar, Kevin
D	85	3	7	Lithograph	Mido (?)	V	450.00	Fonseca, Harry
D	85	3	8	Serigraph		V	275.00	Stroud, Virginia
D	85	3	9	Serigraph		V	275.00	Stroud, Virginia
D	85	3	10	Serigraph		V	250.00	Stroud, Virginia
D	85	3	11	Serigraph		V	250.00	Stroud, Virginia
D	85	3	12	Painting	Comanche	V	3500.00	O'Leary, Diane
D	85	4	1	Jar, Corrugated	Tomque	V		
P	85	4	1	Pot	San Ildefo	V	99.00	

phabetic object name list (see Figure 18 for sample pages). Such a list can be extremely valuable in locating all kinds of errors and problems associated with any word list. In this case, Object Name, Pueblo, and Artist are the three fields where such a listing can be helpful.

The procedures involved in the preparation of an alphabetic word list are: (1) select the fields that you will want to print (with enough room for the maximum field length)—Object Name and the four fields comprising Accession Number;

Figure 18a
Alphabetic Listing of the IPCC Catalog by Object Name ("Adze" to "Ashtray")

1

OBJECT NAME	METHOD OF ACQUISITION	ACCESSION NUMBER		
	L		101	39b
	L		101	39c
	L		101	39d
	L		101	39e
Adze, Basalt	D	78	12	1
Afghan, Crocheted	L	76	51	1
Afghan, Crocheted	L	76	60	1
Album, SW Postcards	D	84	13	1
Album, SW Postcards	D	84	13	2
Apron, Embroidered	L		114	14
Apron, Embroidered	D	76	3	42
Apron, Embroidered	D	76	3	43
Apron, Embroidered	D	76	3	44
Apron, Embroidered	D	76	3	45
Apron, Embroidered	D	76	3	46
Apron, Embroidered	D	76	3	47
Apron, Embroidered	D	76	29	1
Apron, Embroidered	D	78	3	40
Apron, Embroidered	D	78	3	41
Apron, Embroidered	D	78	3	42
Apron, Embroidered	D	78	3	43
Apron, Embroidered	D	78	3	44
Apron, Embroidered	D	78	3	45
Apron, Embroidered	D	78	3	46
Apron, Wedding	D	76	56	1
Arm Band, Bead	L	76	2	3
Arm Band, Bead	L	76	2	4
Arm Band, Beaded	D	78	3	38a
Arm Band, Beaded	D	78	3	38b
Arm Band, Beaded	D	78	3	39a
Arm Band, Beaded	D	78	3	39b
Armband, Bead	L	76	2	10
Armband, Bead	L	76	2	26
Armband, Bead	L	76	2	37
Armband, Bead	D	76	4	15
Arrowhead Fragment	D	78	12	3
Arrowhead Fragments, Quartz	D	78	12	13
Ashtray	D	76	3	32
Ashtray	D	76	10	26
Ashtray	D	76	34	1
Ashtray	P	85	1	62
Ashtray	P	85	1	64
Ashtray	P	85	1	68
Ashtray	P	85	1	86

(2) create the report format template that will be most useful; (3) sort the key field—Object Name—in alphabetic order; and (4) print. Printing is quite rapid on the Imagewriter. Nevertheless, printing 40 pages is by far the slowest part of the entire process.

Figure 18b
Alphabetic Listing of the IPCC Catalog by Object Name ("Oven Paddle" to "Painting")

OBJECT NAME	METHOD OF ACQUISITION	ACCESSION NUMBER		1
Oven Paddle	D	76	3	24c
Oven Stick	D	76	3	24d
Owl	L	76	95	2
Owl, Beaded	L		101	15
Owl, Beaded	L		101	16
Owl, Polychrome	L	76	92	4
Owl, Pottery	D	86	3	4
Painting	D	75	1	1
Painting	D	76	97	8
Painting	P	81	16	19
Painting	P	81	16	26
Painting	P	81	16	28
Painting	P	81	16	34
Painting	P	81	16	41
Painting	P	81	16	42d
Painting	D	82	4	1
Painting	D	83	3	1
Painting	D	85	3	12
Painting, Acryllic	D	77	2	1
Painting, Watercolor	D	76	8	1
Painting, Watercolor	D	77	1	1
Painting, "Apache Fire Dancer"	P	86	5	2
Painting, "Bird"	P	85	9	2
Painting, "Bow Dance"	P	85	6	7
Painting, "Buffalo Dancer"	D	81	8	1
Painting, "Buffalo Dancer"	P	81	16	31
Painting, "Buffalo Dancer"	D	85	9	1
Painting, "Buffalo Hunters"	D	76	72	1
Painting, "Corn Dancer"	P	85	6	6
Painting, "Corn Dance"	D	85	5	2
Painting, "Corn Dance"	P	85	6	1
Painting, "Corn Maiden"	D	83	7	1
Painting, "Dancers of the Rain"	D	77	16	1
Painting, "Deer Dancer"	P	81	16	18
Painting, "Deer Dancer"	P	85	6	3
Painting, "Eagle Dancers"	D	76	7	1
Painting, "Eagle Dancers"	P	78	19	1
Painting, "Eagle Dancer"	P	85	9	1
Painting, "Fruits of Life"	P	78	4	2
Painting, "Harvest Dancers"	P	86	5	1
Painting, "Harvest Man"	D	76	65	1
Painting, "Horse"	D	77	30	1
Painting, "Koshare"	P	85	9	3
Painting, "Matachine"	P	84	8	1
Painting, "Mud Head Dancer"	D	76	79	3
Painting, "Muddy"	P	85	6	2

Figure 18c
Alphabetic Listing of the IPCC Catalog by Object Name ("Vase, Double Spout" to "Wedding Vase")

OBJECT NAME	METHOD OF ACQUISITION	ACCESSION NUMBER		1
Vase, Double Spout	P	85	1	58
Vase, Effigy	P	81	16	1
Vase, Horned	D	77	19	1
Vase, Mica	L	76	116	1
Vase, Polychrome	D	76	51	4
Vase, Polychrome	D	76	54	1
Vase, Polychrome	P	85	1	57
Vase, Polychrome	P	85	1	59
Vase, Polychrome	P	85	1	81
Vase, Polychrome	P	85	1	99
Vase, Polychrome	P	85	1	107
Veil	L		156	4
Veil, Humeral	L		168	3b
Vessel, Cooking	D	82	6	5
Vessel, Cooking	D	82	6	6
Vessel, Double	D	80	1	6
Vest, Embroidered	L	76	6	11
Vest, Embroidered	L	76	29	3
Vest, Embroidered	L	76	108	1
Vest, Leather	L	76	2	28
Vest, Painted	D	76	30	1
Wall Hanging, "Bicentenial Calendar"	D	76	79	5
Wall Hanging, "Calendar"	D	76	79	6
Wallet, Leather	L	76	2	2
War Club	L		132	3
Watchband	D	78	2	13
Watchband	D	78	2	20
Watchband, Silver	D	76	44	96
Watchband, Silver	D	76	44	114
Watchband, Silver	D	78	2	15
Wedding Vase	L		114	3
Wedding Vase	L	76	2	40
Wedding Vase	D	76	10	12
Wedding Vase	D	76	15	2
Wedding Vase	D	76	20	2
Wedding Vase	D	76	21	1
Wedding Vase	D	76	38	1
Wedding Vase	D	76	46	1
Wedding Vase	D	76	74	1
Wedding Vase	D	76	86	1
Wedding Vase	D	79	4	1
Wedding Vase	D	81	15	13
Wedding Vase	P	81	16	2
Wedding Vase	P	81	16	4
Wedding Vase	P	85	1	26
Wedding Vase	P	85	1	56

Many things can be learned from a careful examination of an alphabetic word list. In Figure 18a, we see immediately that we have four records without any entry in the Object Name field, and two records immediately following these that appear to be completely blank. The four turned out to be subrecords for five objects identified only on the first record as ''jewelry,'' and they give us the opportunity to emphasize that *every* record must have something entered in the Object Name and the Accession Number fields. Without this, there is no way of knowing what the object is or of identifying it in a positive (numeric) manner. The blank records were the result of hitting the Return key twice.

Figure 18a also indicates that we were inconsistent in our spelling. Is ''arm-band'' one word or two? The answer is not important, but the consistency is essential (on a further page we found three different spellings for the name of the same object). Obvious typing errors do not appear in Figure 18a, but they do stand out clearly in the midst of the same word spelled differently. For example, on another page (not illustrated and since corrected) there was a series of ''bean pots,'' followed by the word ''bean plot.'' Errors like this are easily caught and flagged for correction in an alphabetic list of object names.

Figure 18b illustrates several word-use problems. The convention those taking the inventory tried to follow was to list the key word for any representational art form (e.g. painting, lithograph, photograph) followed by either the title given by the artist or the subject matter. When taken together with the field for Artist Name, this provides a good identification of these objects. However, note the number of paintings with no information beyond the word ''painting.'' And on three of the records, the inventory taker decided to record not the title or subject but the medium (''watercolor'' on two, ''acrylic'' on one).

As pointed out earlier in this chapter, the utility of a data base is greatly reduced when different kinds of observations are intertwined within the same data field. It is difficult enough to control the consistency of words used in the Object Name field without adding to the confusion by bringing in extraneous information. Note that these observations have nothing to do with the computerization of the museum's artifact records; rather, they have to do with the *usefulness* of the computerized records. The rule that the computer forces us to follow is simple: consistency; consistency; and consistency.

The third page shown from the IPCC list of object names (see Figure 18c) illustrates several additional word conventions. In general, objects such as ''vase'' are further subdivided by appending, after the comma, appropriate words to describe physically observable characteristics of the object. However, ''wedding vase'' is considered a two-word, compound phrase, and thus is alphabetized under ''w'' rather than ''v.'' Similarly, ''wall hanging'' and ''war club'' are word phrases and are not inverted.

There are a number of different problems involved with using words in a consistent manner—miniatures, replicas, parts of objects, and sets of objects, to name but a few. Alphabetized computer listings help to identify the problems, and books such as *Nomenclature for Museum Cataloging* (Chenhall 1978) can aid in resolving them, but each institution must ultimately make its own decisions regarding the conventions that will be followed by recording artifacts and specimens.

PLANS FOR THE FUTURE

Working files such as the IPCC object name listing are the underlying reason for which a computerized museum catalog is created. They are intended to serve the real information needs of the institution, and in form and content are limited only by the imagination of the staff. Once the truly hard work is completed and the collection is recorded in machine-readable form, it is possible to create all kinds of special-purpose reports. The procedure to accomplish this is no more complicated than what has been described here. It is always a simple matter of:

1. SELECTION—of fields of information that may be significant to the problem at hand
2. SORTING—of the selected data in the sequence that will be the most useful
3. PRINTING—the sorted information

Sometimes—and this is the case with the IPCC file—it is only possible to see the potential utility of computer processing after the initial inventory has been recorded, when one belatedly realizes how much more useful it could be, ". . . if only we had also included (thus and so)." When the inventory at the IPCC was recorded, Location was a single-digit code to indicate that the artifact was on display or in storage. Now it is recognized that by expanding this code to perhaps a two- or three-segment composite code, it would be possible not only to select from the file objects that are on display, but to sort the selected records by specific display location and print separate pages for the artifacts that should be in each display. The taking of physical inventories in the future would thus become a simple matter of confirming whether preprinted listings are correct, rather than the cumbersome task of writing everything down and having to go back and compare what is written against the files, item by item.

Fortunately, with the OverVUE program it is relatively easy to add, delete, or change the length of fields. As soon as a system of describing museum location in more precise terms is decided upon, additional fields will be added to the master catalog file to provide for the recording of location within that system. Preprinted listings will be prepared for the next physical inventory from the present file, with spaces for entering just the additional, specific location data. This data will then be added to the master file so that it will be available in the future. The present, simplified location data fields will then be eliminated.

Computerized artifact files are never static. In time, the volume of changes will subside, but as long as the file is being utilized, new ways will be found to expand or change it—perhaps just a little here and there to accomplish this or that purpose, but isn't that what life is all about? Hopefully, the information in this book, including the somewhat detailed description of the IPCC experience, will aid you, the reader, in the development of a viable, computer-oriented, information system for your institution.

Glossary

Data processing terminology is far from uniform. This glossary defines terms as they are used in this book. The reader should understand that others will use the expressions listed here in different ways, that concepts used here are often given other names and, finally, that many individuals who work in this field believe, in all honesty, that their own usage is in some way "standard," all others being incorrect. In their eyes this glossary, like *any* glossary except their own, will appear to be shot through with error.

The present authors make no claim of "authoritative" usage and offer what follows in an effort to render this book accessible to the reader of standard, dictionary English. We hope, too, that with this warning our glossary may at least prove helpful in comprehending other texts that treat data processing, data base management, and museum documentation.

ACCESSION FILE: The cumulative master list of additions to a collection, by transaction and lot; the foundation of all collection documentation. *See also* Decession; Lot

ANALOG: The representation of information by a continuous wave form such as the groove in a phonograph record (one-dimensional) or the varying density of a photographic negative (two-dimensional); so called because analog information is often an "imprint" of, hence analogous to, the form recorded. In electronic systems analog information is often represented by a continuously

Figure 19
Examples of Arrays

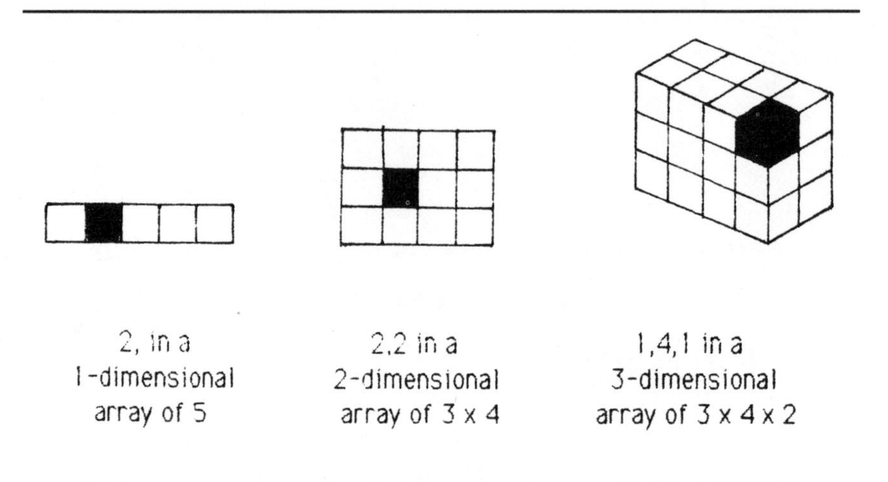

varying voltage, as opposed to the discrete, timed cycles of a digital computer. *See also* Digital

APPEND: To add new logical records at the end of an item file or new paragraphs at the end of a text file. One of the ways to "update" a file. *See also* Item File; Logical File; Paragraph; Text File; Update

APPLICATIONS PROGRAM or APPLICATIONS SOFTWARE: A relative term, implying specificity of function. It excludes operating systems and everything else regarded as "systems software" and includes all user-written programs and procedures. Between those extremes, what is an applications program depends upon the speaker's point of view. *See also* Program; Procedure; Software

ARCHIVE: To write data to archive storage. *See also* Storage; Write

ARRAY: A group of comparable data values, so organized as to represent a regular, rectilinear arrangement in space. Each element (individual value) in an array is designated by its position in each dimension (see Figure 19). A one-dimensional array is the same as a list. *See also* Value

ASCII (*American Standard Code for Information Interchange*): One of two common conventions associating keyboard characters with specific bytes, by number. Since alphanumeric sorting is done by byte number, the choice between ASCII and EBCDIC influences the sorting sequence, e.g. whether numerals come before or after letters. ASCII is used in nearly all microcomputer systems and most non-IBM minicomputers and mainframes. *See also* Byte; Character; EBCDIC; Sort; Word

ASSEMBLE, ASSEMBLER, ASSEMBLY: *See* Programming Language

Figure 20
Decimal Numbers with Binary Equivalents

Decimal number	Binary Equivalent
.5	.1
0.	0.
1.	1.
2. *	10.
4. *	100.
8. *	1000.
10.	1010.
15.	1111.
16. *	10000.
100.	1100100.

Numbers marked with an asterisk are ''round numbers'' in binary arithmetic

AUTHORITY LIST, AUTHORITY FILE: A list of standardized terms or values that may be used in a given field or domain, all other values being ''illegal'' in that field or domain. An authority file contains an authority list. Cf.: Domain; Field; Value

BACKUP: Copying files, including software, for off-line, archival storage as a precaution against the inevitable loss of files on magnetic tape or disk. An essential step whenever a permanent file is created or updated. *See also* File; Generation (usage 4); Storage

BATCH PROCESSING: Automatic processing when all commands are prepared in advance and there is no opportunity for user intervention during processing. Batch processing was once the norm (when all input consisted of punched cards); it remains useful for time-consuming operations such as searching or sorting very large files. *See also* Command

BINARY NUMBER: Any number expressed with only two value symbols, ''0'' and ''1'' rather than ten or any other number of digits. Positions to the left of the radix point have values corresponding to powers of two: units, twos, fours, eights, etc., rather than units, tens, hundreds, etc., as with decimal numbers (see Figure 20). Computers use binary arithmetic because digits must be represented by electronic switches, which have only two positions, on and off.

BINARY SEARCH: A common strategy for finding an item in an ordered list such as a lexical index, a sorted file, or a telephone book. You look near the middle to determine in which half the item must be, then near the middle of one half and so on until you close in upon the location where the desired item must be, if it is present. The list must be ordered upon the field of the item being sought, thus you can use it to find a name in a telephone book, but not to find out who has a particular number. Cf.: Field; Lexical Index; Sort

BIT: The smallest unit of information in electronic systems, represented by switches that are on or off, voltages that are high or low, or small areas of material in one of two possible magnetic states. A bit may be interpreted as a binary zero or one, yes or no, true or false, stop or go, or any such dichotomy. More complex information is represented by a series or ''strings'' of bits. The smallest such group is called a byte. *See also* Binary Number; Byte; Word

BIT-MAPPED SCREEN: A screen display comprising an array of individually controlled pixels and capable of displaying any shape. The screen is associated with a block of dedicated main storage—the bit map—in which one or more bits control each pixel. Characters are generated by software rather than a hard-wired character generator. *See also* Array; Bit; Character (usage 1); Pixel; Storage (usage 1a)

BLOCK, BLOCKING FACTOR: The amount of data copied to main storage or to cache storage in one file access. When a block is defined as a certain number of physical records, this number is called a blocking factor. Among microcomputer systems a block is often one track sector. *See also* File Access; Physical Record; Sector; Storage

BOOLEAN OPERATOR, BOOLEAN QUERY: *See* Query

BRANCHING FACTOR: In a regular tree structure, the number of branches descending from each node except the leaves. A tree with a branching factor of two is called a binary tree. *See also* Tree

BYTE: A series of bits stored, moved, and processed as a unit. All computer systems process information by the byte, and in any one system all bytes are the same size, being six, seven, eight, or nine bits. The eight-bit byte is most prevalent. Since there are 256 possible combinations of eight zeros and ones, an eight-bit byte allows the use of up to 256 letters, numerals, and special characters. Small groups of bytes processed as a unit are, unfortunately, called ''words.'' *See also* Bit; Word

CACHE: *See* Storage

CATALOG: (1) A file describing individual objects or a set or series of related objects that is organized in a meaningful way so that like things are together insofar as possible. A computerized catalog must be keyed to the accession file by reference. (2) A publication containing a catalog list. It is usual to include some accession-file data in such a list, since that file may not be available to readers. *See also* Accession File; File; Key

CATHODE RAY TUBE: A kind of vacuum tube used as the screen of a television receiver or monitor and as the display screen of most computer terminals and microcomputers. Often called a CRT. Devices that incorporate a CRT are called video display terminals or VDTs.

Figure 21
Examples of Coding

Literal Term	Code Word	Code Number
Bequest	B	1
Collected in the field	C	2
Gift	G	3
Loan	L	4
Purchase	P	5
Source unknown	U	6

CAT-SCAN: CAT stands for "computer-assisted tomography," a technique useful for imaging the hidden interior structure of an object, such as a mummy. X-rays or other penetrating radiation pass through the object from all directions and a computer is used to average the blockage of radiation for every interior point or voxel. The greater the average blockage, the greater the density of the voxel. *See also* Voxel.

CENTRAL PROCESSING UNIT: *See* Processing Unit

CHARACTER: (1) A letter, numeral, punctuation mark, or other symbol, including the blank space produced by the space bar; (2) Keyboard characters refer to all of the above plus "control" bytes that signal such functions as backspace, tab, carriage return, page feed, and end of file; (3) Any byte, including all of the above plus nonprinting characters not associated with any keystroke; (4) A data type also called "text." Fields of this type may accept as input only the characters of definitions (1) and (2), but in some systems accept any byte as character data. In general, character fields may only be processed by editing commands, not arithmetically, and they are sorted alphabetically. *See also* Byte; Data Type; Field

CHARACTER SET: A closed list of characters or bytes, often associated with data definitions and data types. Thus a character or text field may (or may not) be restricted to the approximately 95 keyboard characters of the ASCII character set. A number field may accept only the set "+,−,.,0,1,2,3,4,5,6,7,8,9," all other input being rejected as invalid. A flag field may be restricted to the set "Y" or "N". *See also* ASCII; Byte; Character; Data definition; Data Type; Field; Flag Field; Text

CODING: (1) Writing a computer program at any level; (2) Substitution of standardized code words or numbers for literal terms in a file to save space and achieve uniformity of length (see Figure 21). Coding is not to be confused with encryption. *See also* Encryption; File; Length; Programming Language

COLLISION: *See* Hashing

COLUMN: *See* Table

COMMAND: An input expression, usually from the keyboard, that initiates specified processing. It must be given when software requests a command word (the name of a program, module, or procedure), usually followed by a "parameter list" of more detailed specifications. For example, a FIND command must be followed by specification of what is to be found and what files are to be searched. A *command sequence* is a procedure that may be named and stored in a procedure file as a *macro statement,* which may then be used itself as a command to initiate the entire sequence. *See also* File; Module; Procedure; Procedure File; Program; Prompt

COMPILE, COMPILER: *See* Programming Language

COMPOSITE FIELD: A field made up of two or more subfields. A repeatable or multi-valued composite field is the same as a repeating group. *See also* Field; Repeatable Field; Repeating Group; Subfield

COMPOSITE KEY: A key comprising more than one field or domain. For example, in the IPCC data base discussed in Chapter 13, four fields combine to form the accession number data that uniquely identifies a record and serves as its primary key. These four fields are a composite key. Note, too, that the first three form a composite secondary key, as do the first two. *See also* Domain; Field; Key; Record

COMPRESSION: *See* Data Compression

COMPUTED FIELD: A field in which the value is not input directly or copied but rather computed from the value of one or more other fields, e.g. a measurement in centimenters computed from one in inches or "artist's age" computed from "year of birth" and "date of origin." *See also* Field; Input; Value

CONCATENATE: To combine two or more units of data end-to-end, without changing them in any other way. For example, if two files of records in alphabetical order were concatenated, the resulting file would contain all records of the first file from A to Z, followed by all those from the second from A to Z. *See also* File; Merge; Record

CONTROL FILE: A file maintained largely or entirely by the system for its own use, e.g. to remember the user's naming and description of files, record formats, data fields, and the physical location of logical files. *See also* Field; File; Logical File; Record Format

CONVERSION: In data processing, the substitution of one data type for another with no change in meaning or loss of information, e.g. from character representation to short integer or vice versa. Data from the keyboard are always in character representation, one byte per keystroke; the data may be stored as it is or converted to a numeric data type such as short integrer, using exactly two bytes per number. The bytes are different, the sense the same. *See also* Byte; Character; Data Type; Integer; Storage

COPY PROTECTION: Any provision built into software or a data base to prevent unauthorized use or dissemination. In the extreme case, copy protection prevents installation of software on a user's own hard disk, so that the system can only be used with the supplier's floppy disk mounted. More often the user may "install" or copy the software once but not a second time without a "deinstallation" process that disables the hard disk copy before reactivating the supplier's disk. Not all commercial software has copy protection. *See also* Data Base; Software

CORE: An obsolete technology for main storage after vacuum tubes and before integrated circuitry. The term is still used sometimes for "main storage," especially in the expression "core dump." *See also* Dump; Storage

CPU: *See* Processing Unit

CRT: *See* Cathode Ray Tube

CURSOR: A position marker on a video display terminal, often in the shape of an underscore or an arrow, about the size of a character and often flashing for greater visibility. The cursor moves independently of other displayed data, sometimes under program control and sometimes under user control. It may indicate the position of the next character to be input, the user's selection from a menu, the movement of data or functions into a program's window, or the locus of an outline image during input. Cursor motion may be controlled by a mouse. *See also* Cathode Ray Tube; Function; Menu; Mouse; Option; Outline Image; Window

DATA BASE: One or more files serving a common purpose, all accessible to the same hardware and software systems at the same time so that data from more than one component file may be combined in a single report. *See also* File; Hardware; Report; Software

DATA BASE MANAGEMENT SYSTEM (DBMS): A coordinated set of programs that, in cooperation with an operating system, controls the input, upkeep, and output of a data base. *See also* Data Base; Input; Operating System; Output; Program

DATA COMPRESSION: Any processing that reduces the number of bytes to be stored or transmitted without loss of information. Compression takes advantage of regularities, such as repetition, in the data and is especially applicable to black-and-white images, which often contain long series of identical pixels and may be compressed by more than 90%. Compressed data must be decompressed for output. *See also* Byte; Output; Pixel

DATA DEFINITION: A set of rules governing a specific field, subfield, or domain, specifying such characteristics as data type, length, precision, syntax, authority list, and repeatability. A "global" data definition resides in a data dictionary and is called a domain. A local data definition, associated with a single file, is called a field description. *See also* Authority List; Data Dictionary; Data Type; Domain; Field; File; Global; Length; Precision; Repeatable Field; Subfield; Syntax

DATA DEPENDENCY: Discussed under "Flat Files and Relations" in Chapter 4

DATA DICTIONARY: A file or set of files containing data definitions that apply not to a specific file but rather to all files of a data base. Such definitions are called global definitions or "domains." Each domain is named but, in many systems, fields in various files need not have the same name as the domain to which they belong. *See also* Data Base; Data Definition; Domain; Field; File; Global

DATA FILE: A file used as a permanent repository of the user's data. Data file contents come mostly or entirely from input and are not copied, selected, or computed from other files. Data files are also called truth files and master files; their content is primitive data. Museum examples include accession, inventory, catalog, and bibliography files—the foundations of museum documentation. *See also* File; Selection

DATA PATH: The series of steps leading from a query or request for data—by way of sequential search, binary search, or an index search—to retrieval of the data or a determination that it is not available. Each step may involve a file access, but one access will sometimes advance the search more than one step along the data path. *See also* Binary Search; File Access; Index; Query; Sequential Search

DATA TYPE: A classification assigned to every field and domain that determines how the field content or "value" is stored, hence how it can be processed and, by extension, what data may be entered. Systems vary in the names and definitions of data types recognized. Some common types include character, flag, and integer data, though each type may be known by other names in any given system's documentation. *See also* Character; Domain; Field; Flag; Integer; Value

DATE FIELD: (1) A field or domain for any kind of specific time reference other than time of day; (2) A synthetic data type supported by business-oriented DBMSs. In museum applications outside the business office this type is useless and dangerous, and should be avoided. *See also* Data Type; Domain; Field

DBMS: *See* Data Base Management System

DEACCESSION: The term "decession" is preferred

DECESSION FILE: A cumulative list of objects removed from the collection, by accession number. Although a decession file corresponds in a way to the accession file, recording the end of institutional accountability for listed objects, decessions do not correspond to accessions one-for-one because objects are seldom if ever decessioned in the same lots in which they were acquired. *See also* Accession File; Lot

DECIMAL: A common data type for fixed-point numbers; field length is specified in terms of decimal digits, input and output are in decimal digits, binary arithmetic operations are hidden from the user, and rounded results are correct for decimal arithmetic. *See also* Binary Number; Data Type; Fixed-Point Number; Input; Length; Output

DEFAULT: Any software action taken automatically, where the user might intervene but does not. For example, an input program might enter "unknown," "NA," or "anonymous" by default in fields where the user types nothing. *See also* Field; Input; Option

DEINSTALLATION: *See* Copy Protection

DELETE: To remove any body of data, from an entire software system down to a subfield. Deletion usually does not destroy data but only renders it inaccessible by changing directories and tables of contents, releasing the space for other use. Some systems have an "undelete" function for recovery of deleted parts, as long as the space has not been reused. *See also* Directory; Software; Subfield; Table of Contents; Update

DERIVED DATA: Data not recorded from direct input but obtained from other data by copying, selection, recombination, computation, or any combination of such processing. The content of working files and of reports. *See also* Report; Selection; Working File.

DIGITAL: The representation of information by a sequence of value symbols (digits) such as "0, 1, 2, . . . 9" or, in computer systems, the two bit values "0" and "1". As opposed to analog systems, which represent information by wave forms. In principle any information can be converted from analog to digital representation and vice versa. *See also* Analog; Bit

DIRECT ACCESS: The ability to locate and retrieve physical records by storage address or record number rather than by sequential search, analogous to finding printed information by page reference. Data storage systems, such as disk drives, that support direct access are called Direct Access Storage Devices or DASD, pronounced "DAZ-dee." *See also* Physical Record; Sequential Searching (usage 1)

DIRECTORY: A file or data base maintained automatically by the operating system, containing the names, locations, and descriptions of all other files known to the system. Every search for data begins with a search of the directory to locate files containing the required data. *See also* Data Base; File; Operating System

DISKETTE: A demountable floppy disk of relatively small diameter. The usual medium for dissemination of microcomputer software. *See also* Floppy Disk; Software

DOCUMENT, DOCUMENT FILE: A text file without internal subdivisions except, perhaps, division into lines for display on screen or paper. *See also* Line; Text File

DOMAIN: A global data definition stored in a data dictionary. Each domain has a name, which must be unique among the domains of its data base. *See also* Data Base; Data Definition; Data Dictionary; Global

DONOR FILE: A museum's list of those who have given objects for the collection. The list may or may not include legators, donors of purchase funds, corporate donors, etc. Comparable to a vendor file (usage 2). *See also* Vendor File (usage 2)

DUMP: A printout of data exactly or very nearly as it is stored. Used primarily for diagnosis and trouble shooting, dumps may be difficult to read. A "core dump" represents main storage at a particular instant in time, a file dump the contents of a file, including bytes meant only for system use. *See also* Byte; Storage; Report

EBCDIC (E*xtended* B*inary-Coded* D*ecimal* I*nterchange* C*ode*): One of two common conventions associating keyboard characters with specific bytes, by number. Since alphanumeric sorting is done by byte number, the choice between EBCDIC and ASCII influences the sorting sequence, e.g. whether numerals come before or after letters. EBCDIC is characteristic of IBM and IBM-compatible mainframes and minicomputers. Pronounced "EB-si-dik." *See also* ASCII; Byte; Character; Sort; Word

ENCRYPTION: Systematic transformation of data to prevent unauthorized access or, more precisely, to prevent understanding in case of unauthorized access. Not to be confused with coding. *See also* Coding

FACSIMILE TRANSMISSION: The process of transmitting an image, often of a paper document, pixel-by-pixel from a scanner to a remote printer, where output may be concurrent with scanning. *See also* Output; Pixel; Scanning (usage 2)

FALSE DROP: Retrieval of unwanted data because of (1) homonyms, e.g. "vessel" meaning a ship and "vessel" meaning a container, (2) terms with meaningful substrings, e.g. "silk screen," containing "silk" and "screen," or (3) terms with different meanings in different contexts, e.g. "Washington," one of several cities by that name, or "Washington," one of several persons by that last or first name. False drops occur most often in text searching, where the context is difficult to recognize. *See also* Free Text; Retrieve; Text File

FIELD: A division of a logical record. Just as a logical record is usually devoted to one subject, such as an accession, each field is assigned to one aspect of the subject. e.g. "date of acquisition." The fields of a record format are often named and individually defined as to data type, syntax, vocabulary, and sometimes length. Often, every record of a logical file contains the same set of fields in the same order. *See also* Length; Logical File; Logical Record; Record Format; Syntax

FIELD DESCRIPTION: A data definition applicable to only one field or subfield of one file. Where there is a data dictionary, a field is normally described by reference to the domain it represents. *See also* Data Definition; Data Dictionary; Domain; Field; File; Subfield

FIELD INDEXING: Indexing using entire values of fields or subfields as subject headings, not words or phrases contained within fields. In field indexing, the field must be specified by name. Either the entire index file refers to one field or a field name must be incorporated into each subject heading, e.g. "Matisse, Henri/attribution" rather than "Matisse, Henri." *See also* Column; Domain; Field; Index; Value; Word Indexing

FIELD TYPE: *See* Data Type

FILE: A body of stored data. Each file has a name, a physical storage location, and a formal description. The name must be unique within the storage system where the file resides; the operating system maintains a directory or file of file names, in which every name is associated with a storage location and description. The description enables software to interpret the file's contents correctly. The term "file" may refer either to a logical file or to a physical file. *See also* Directory; Logical File; Operating System; Physical File; Software

FILE ACCESS: An I/O operation copying data from internal storage to external storage or vice versa. Also called Data Access. *See also* Read; Storage; Write

FILE MANAGER: (1) A person responsible for allocating external storage facilities, seeing that files are backed up regularly and archived and deleted when appropriate; (2) Software associated with an operating system or DBMS that creates, stores, and deletes files, assigns physical storage locations, maintains a directory, puts logical records into the required physical format, and sometimes handles automatic indexing. The file manager may be a separate software package from the operating system and DBMS; (3) *The File Manager,* a public domain DBMS in the MUMPS language. *See also* Backup; Data Base Management System; Directory; File; Index; Logical Record; Operating System; Software; Storage

FIXED DISK: A magnetic disk permanently built into its drive. An external storage medium, not usable for archive storage or backup. *See also* Backup; Storage

FIXED FORMAT: A record format requiring that every record of the file contain exactly the same set of fields as every other, in the same order, and that each field be present exactly once (or a fixed number of times) in every record. The majority of all files have fixed formats. *See also* Field; File; Format; Record

FIXED-POINT NUMBER: A common data type for numbers, where the data definition specifies a decimal point a certain number of digits from the right. Such a field or domain might be defined as "fixed 5, 2" or "decimal 5, 2," meaning that the maximum length of the field is five digits and the last two are understood to be to the right of the decimal point. Thus a stored number 12345 would always be treated and displayed as 123.45. A fixed-point number with zero decimal places is the same as an integer; therefore, some systems do not support a separate integer data type. *See also* Data Definition; Data Type; Decimal; Domain; Field; Floating-Point Number; Integer; Length

FLAG, FLAG FIELD: A common data type, also called "logical data," where a field or domain may represent only one of two values, i.e., on/off, yes/no, corresponding to the values of a single bit. *See also* Bit; Domain; Field; Value

FLAT FILE: A fixed format file with the added requirement that each field be present once and only once in every record. *See also* Field; File; Fixed Format; Record

FLOATING-POINT NUMBER: A data type, also called "real number," corresponding to scientific notation. Each number is represented by a pair of values called the characteristic and the mantissa, the latter indicating the position of the decimal point. For example, the decimal number 0.000123 would be 123E-6 in scientific notation, where 123 is the characteristic and -6 the mantissa. A floating-point number usually occupies about six bytes of storage. Many DBMSs do not support this type and there is little use for it in collection management. *See also* Byte; Data Type; Decimal Number; Fixed-Point Number; Value

FLOPPY DISK: A magnetic disk made of plastic, characterized by relatively low data density and storage capacity and relatively slow operation as compared to hard disks. The magnetic reading/writing head rests directly upon the floppy disk surface. Floppy disks are usually demountable, hence used to transfer software and data from one system to another and for archive storage. *See also* Hard Disk; Software; Storage

FORMS DESIGN: A facility of some DBMSs and Fourth Generation Languages for developing and storing procedures enabling a Report Generator to produce a standard report on demand, e.g. a monthly insurance report or a loan receipt. *See also* Procedure (usages 2 and 3); Programming Language; Report; Report Generator

FOURTH-GENERATION LANGUAGE: *See* Programming Language

FREE TEXT: A text file or a field of the text or character type, where the length is variable and the maximum length large or unlimited. *See also* Character; Data Type; Field; Length; Text File; Variable Length

FREQUENCY DICTIONARY: A technique for analysis of the content of an item file and detection of variant spelling and usage. For each field or column

Figure 22
An Excerpt from a Frequency Dictionary

Object Name	Frequency	Record Numbers
water color	3	123,444,780
water-color	5	56,445,890,891,892
watercolor	2134	6,7,8,13,14,124,125,126,127.......
watercolour	1	432
witercolor	1	9

a sorted list of all field values is produced, each value accompanied by a count of its occurrences and, often, a list of the logical records in which the value is found (see Figure 22). *See also* Column; Field; Item File; Logical Record; Value

FUNCTION: (1) A mathematical function: one value depending upon one or more others. For example, "interest" is a function of "balance" and "interest rate"; (2) A comparable nonmathematical data dependence. For example, "date of birth" is a function of "name"; (3) In computer programming, a module that computes one value as a function of one or more others. For example, a square root function (usually named SQRT) given the value "16" computes and returns the value "4." *See also* Data Dependency; Module (usage 2); Value

GENERATION: (1) *System* generation: On large-scale systems, the process of customizing systems software, such as an operating system, by loading a main component plus selected components to meet the requirements of the installation; (2) *Hardware* generations: first generation = vacuum tube technology (now obsolete); second generation = individual transistor technology (also obsolete); third generation = integrated electronic technology, including multiple components such as transistors etched into silicon chips; (3) *Software* generations: first generation = machine language; second generation = assembly and macro languages; third generation = high-level languages; fourth generation = nonprocedural languages; fifth generation is not yet well defined; (4) *File* generations: generation zero = the current state of a file; generation one = the state of the file when backed up, prior to the most recent updating; generation two = the state of the file when backed up prior to generation one, etc. *See also*

Backup; File; Fourth Generation Language; Hardware; Microchip; Operating System; Software; Storage

GIGA-: *See* Kilo-

GLOBAL: Applicable throughout something, as a global data definition or domain applicable throughout an entire data base. *See also* Data Base; Data Definition; Domain

GLOBAL UPDATE: A process that replaces one field value with another throughout an entire file, e.g. replacing ''Republic of Congo'' with ''Zaire'' in the country of origin field throughout an object catalog. One of the ways to update a file. *See also* Field; File; Update; Value

GRIPHOS: A batch-processing data base management system developed for the United Nations Library in the late 1960's and adapted for museum applications in collaboration with the Museum Computer Network. Now obsolete and unavailable. *See also* Batch Processing; Data Base Management System

HARD COPY: Output recorded on a lasting medium such as paper or microfilm, as opposed to display on a screen. *See also* Output

HARD DISK: A magnetic disk made of metal, characterized by relatively high data density and storage capacity and relatively fast operation as compared to floppy disks. The magnetic reading/writing head never makes physical contact with the hard disk surface. *See also* Floppy Disk

HARDWARE: All tangible components of a data processing system, including peripheral input, output, storage, and communications devices, but sometimes not such ''data media'' as punched cards and tape, magnetic tape, and moveable disks. Built-in tapes and disks are hardware components. *See also* Input; Output; Software; Storage

HASHING: An indexing strategy in which every key to be indexed is converted to a number by computation based upon the numerical values of its component bytes, with this number used as the storage location of the index entry. Keys used in queries are subjected to the same computation so that the search goes directly to appropriate index entries. Hashing results in the shortest possible data path; however, it is subject to ''collisions'' when two or more keys yield the same storage location, and it is almost useless for range queries. *See also* Byte; Data Path; Index; Index Entry; Key; Query

HELP: An almost universal command, often shortened to ''h,'' included on most menus. This command interrupts processing to display textual information about the software or module that is active, including options available to the user at the point of interruption. Often the information offered is drawn from the system manual. *See also* Command; Menu; Module; Option; Software

HIGH-LEVEL LANGUAGE: *See* Programming language

HOLOGRAPHY: Discussed under ''Image Capture'' in Chapter 9.

INDEX: A file used to locate information in other files. An index is to a data base largely what a book's index is to its content. *See also* Data Base; File; Inverted Index

INDEX ENTRY: One unit or logical record in an index file. It is comparable to an entry in a book index, containing one subject associated with a list of physical locations where data about the subject may be found. *See also* Index; Logical Record

INPUT: (1) The passage of information from any external source, such as a keyboard, digitizer, card reader, or automatic sensor into an electronic data processing system. Also the entry of information previously output by the same system or another onto a tape, moveable disk, or communications line; (2) That which is input. *See also* Output

INSERT: To add a new logical record or paragraph to a file in an ordered position, not simply at the end, or to add a field or subfield to a logical record. One of the ways to update a file. *See also* Field; File; Logical Record; Subfield; Update

INSTALL: *See* Copy Protection

INSTRUCTION: (1) One step in a machine-language program, consisting of a few binary numbers, the first of which specifies an instruction (usage 2) to be executed; (2) One numbered component of a processing unit, e.g. the "add" instruction or the "move character" instruction. *See also* Binary Number; Machine Language; Program; Processing Unit

INTEGER: (1) A whole number, positive or negative, but without a decimal point. The point is presumed to be immediately to the right of the last digit. Integers can be stored as character data in some systems, but must always be converted to the integer data type before an arithmetic operation; (2) The integer data type containing binary numbers suitable for arithmetic operations. Many systems support both "short" and "long" integer types. In systems using the eight-bit byte, a short integer is two bytes or 16 bits, with a range from −32,768 through 32,767, while a long integer is four bytes, with a range from −2,147,483,648 through 2,147,483,647, zero being considered a positive integer. Some mainframe systems support larger integers. *See also* Binary; Bit; Byte; Character; Conversion; Data Type

INTERPRET: *See* Programming Language

INVENTORY: A list of physical objects for which an institution is accountable. In this book, usually a location inventory showing the current location and/or the "home" location of each item, or a tracking inventory, which includes a cumulative history of the object's moves. A single catalog entry may require a number of inventory records if the object is a set or series or an object with parts that may be separated for storage or display. *See also* Catalog

INVERTED INDEX: An ordered list of keys or keywords, each associated with a list of occurrences. Analogous to a book index. References may be pointers to physical storage locations or may refer to logical records identified by primary key. *See also* Index; Key; Keyword; Logical Record; Pointer

I/O: Conventional abbreviation for "input and output." I/O operations copy data to and from internal storage. All external storage I/O involving input and output devices entails mechanical motion, hence delay in processing. Exceptions are output to a screen display and input from some automatic sensing devices. Processing that requires extensive I/O is said to be "I/O bound." Unfortunately, this includes much data base processing. *See also* Data Base; Input; Output; Storage

ITEM FILE: Any file divided into records with a repeating format, each record holding the same set of data fields as every other, but describing a separate entity. Item files are by far the most common kind. They include fixed format files, flat files, and tables, but not unstructured text files. Examples include accession files, shelf lists, and mailing lists. *See also* Field; File; Fixed Format; Flat File; Format; Record; Table; Text File

JOIN: A relational data base operation where fields, rather than logical records, from two or more files or relations, are recombined to form a new relation. The new relation is organized about a key that is common to all the original relations. Discussed under "Combining Files" in Chapter 7. *See also* Field; File; Key; Logical Record; Relation; Relational

KEY: A field or set of related fields that may be used to sort or index records. Most fields other than free text may be used as keys. Keys are either primary or secondary. A primary key must have a different value in every record, e.g. the accession number field(s) in an accession file. A secondary key may have the same value in more than one record, e.g. the source field in an accession file. A secondary key classifies the record in which it is found. *See also* Field; File; Record; Value

KEY FIELD:—See Key

KEYWORD: A word or phrase deliberately used in a text file or a text field to force retrieval of the record when the file is searched for the keyword. Ideally, keywords, in standard spelling, are selected from an authority list. Not to be confused with a key field. *See also* Authority List; Record; Text File; Text Field

KILO-: A numeric prefix (see Figure 23)

KWIC (K*ey* W*ord* i*n* C*ontext*): A technique for text analysis. A list of excerpts is produced, one for each word of text (sometimes excluding articles and prepositions). In each excerpt the word is accompanied by a few words of the preceding and following text. All excerpts are displayed as a list, sorted on the word and the following (not the preceding) context. A sample KWIC output from this paragraph is shown as Figure 24. *See also* Sort; Text

Figure 23
Numeric Prefixes

Prefix	Standard Meaning	Meaning in Data Processing	Abbre-viation	Examples
Kilo-	one thousand (10^3)	1,024 (2^{10})	K	Kilobyte, KB
Mega-	one million (10^6)	1,048,576(2^{20})	M	Megabyte, MB
Giga-	one billion (10^9)*	1,073,741,824(2^{30})	G	Gigabyte,GB
Tera-	one trillion(10^{12})	1,099,511,627,776 (2^{40})	T	Terabyte,TB
Milli-	one thousandth(10^{-3})	1/1024(2^{-10})	m	Millisecond
Micro-	one millionth(10^{-6})	1/1,048,576(2^{-20})	μ**	Microsecond
Nano-	one billionth(10^{-9})*	1/1,073,741,824 (2^{-30})	n	Nanosecond
Pico-	one trillionth(10^{-12})	1/1,099,511,627,776 (2^{-40})	p	Picosecond

* U. S. usage. In British usage one billion is equal to what, in the U. S., is called one trillion. ** This is the Greek letter *mu*.

LANGUAGE: (1) A programming language; (2) Any set of rules for preparing information to guide data processing. Examples include query languages to specify data retrieval, data definition languages to define fields and domains, and format description languages to specify the layout of reports. *See also* Data Definition; Programming Language; Query; Report

LENGTH: The size of anything in storage, expressed in smaller units, i.e., the length of a file in bytes, lines, pages, or physical records, the length of a record in bytes or lines, and, especially the length of a field in bytes. For

Figure 24
KWIC (*Key Word in Context*) Analysis

line 4	and following	text	All excerpts are
line 1	technique for	text	analysis. A
line 2	each word of	text	(sometimes excluding
line 4	sorted on the	word	and the following
line 3	the excerpt the	word	is accompanied
line 2	one for each	word	of text is
line 3	by a few	words	of the preceding

variable fields a distinction is often made between actual length (space used) and maximum length (space available). The length of a numerical field is a measure of its precision. The term "field width" is interchangeable with "field length." *See also* Byte; Field; Line; Physical Record; Precision; Variable Field

LEXICAL INDEX: An index in which all the entries are listed in alphabetic or numeric order, as in a book's index or a telephone book. It is suitable for binary searching. *See also* Binary Search; Index; Inverted Index

LINE: In text files, including program source code, either the physical records, the logical records, or both may be called lines. This usually but not necessarily implies records short enough to be displayed as a single line on a screen or printed as a single line on paper. *See also* Logical Record; Physical Record; Source Code; Text File

LINKED RECORD: A record containing specific references or "pointers" to related records so they may be retrieved at the same time, when that is appropriate. For example, the record of an accessioned lot may contain "down pointers" to object records of individual items, while an object record may contain an "up pointer" to the accession record. *See also* Pointer; Record

LIST PROCESSING: (1) In artificial intelligence and the LISP programming language, a software design that eliminates all length restrictions, at some cost in efficiency. Also implemented in hardware as the "LISP machine." This is the original meaning. (2) In DBMSs, software for processing repeating and repeatable fields; (3) In indexing, software for processing the list of physical record locations contained in an index entry; (4) In word processing, software for generating such items as form letters from a single text and a separate list of names and addresses. *See also* Data Base Management Systems; Index Entry; Length; Repeatable Field; Repeating Group

LISTING: Formerly the same as "printout," but now sometimes also a screen display. This term does not imply an actual list of anything. *See also* Report

LOGICAL DATA: *See* Flag

LOGICAL FILE: A body of data stored in one or more physical files. Subdivisions of a logical file are logical records, which may or may not coincide with physical records and may or may not be uniform in length. The relationship between logical files and physical files is analogous to the relationship between the content of a book and the volume(s) on which it is printed. In modern systems the user is concerned almost exclusively with logical files, leaving the management of physical files to software. *See also* Logical Record; Physical File; Physical Record; Software

LOGICAL RECORD: A subdivision of a logical file. In a given file, logical records may or may not be of uniform length and may or may not coincide with physical records. Their relationship to physical records is analogous to the relationship of paragraphs to pages. In item files, each logical record is often devoted to one subject, e.g. one accession record. *See also* Item File; Logical File; Physical Record

LOT: One or more objects acquired in a single transaction, as by bequest, field collection, gift, or purchase, and covered by a single record in the accession file. *See also* Accession File; Record

MACHINE LANGUAGE: The form of a program reduced to binary numbers representing a series of instructions to be executed in order. Each processor has its own set of instructions and built-in rules for decoding machine language. Hence machine languages are also called "native code." *See also* Binary Number; Instruction; Processor; Program

MACRO, MACRO FILE: *See* Procedure (usages 2 and 3)

MACRO LANGUAGE, MACRO STATEMENT: *See* Programming Language

MASS FIELD UPDATE:—See Global Update

MASTER FILE: (1) A data file; (2) A directory; (3) Any file that makes reference to other files for further information. The term is sometimes used loosely and variously. *See also* Data File; Directory; File

MEGA-: *See* Kilo-

MEMORY: The term "storage" is preferred. *See also* Storage

MENU: A numbered list of options, such as commands, displayed on a screen so that the user may select one by merely typing its number or moving the cursor. If the chosen command requires a parameter list, the menu may be replaced by a sequence of additional menus of choices. One common option in every context is "help," which evokes a display of explanatory material, often from the system manual. *See also* Command; Cursor; Option

MERGE: To combine the content of two or more files sorted in the same order in such a way that all records remain intact but are interleaved to maintain the sorted order. *See also* Concatenate; File; Record; Sort

MICRO-: *See* Kilo-

MICROCHIP: A miniaturized component of a computer or other "intelligent" hardware, today usually about the size of a fingernail and composed primarily of crystallized silicon, though other crystals can be used. Microchips include processing units, main storage (RAM), and chips supporting special uses such as communications and color display. *See also* Main Storage; Processing Unit

MILLI-: *See* Kilo-

MODEM: Short for *mo*dulator-*dem*odulator. A hardware device that modulates outgoing signals for long distance transmission and demodulates incoming signals for computer processing. A modulated signal is analog and serial, carried by an alternating current of moderate voltage. A demodulated signal is digital and parallel, carried by direct current of very low voltage. *See also* Analog; Digital

MODULATE: *See* Modem

MODULE: (1) A program file; (2) A program, but usually a small program invoked by other programs to perform some repetitive, standardized task. Software such as a DBMS or an operating system usually includes a "library" of program modules at the service of the main programs. *See also* Data Base Management System; Operating System; Program; Program File; Software

MOUSE: A device used in conjunction with a video display terminal, consisting of a hand-held vehicle moving over a flat surface near the terminal. The mouse senses the direction and distance of its own movements, which appear as corresponding motions of a cursor on the screen. Motions of the cursor may be used to draw or trace images, select options from a menu on the screen, and move or "drag" data and functions from one screen window to another where such movements represent communication between computer processes, for example, association of a file with a running program, association of a type font with a body of text, or association of a color with an imaged shape. *See also* Cathode Ray Tube; Cursor; Function; Menu; Option; Outline Image; Window

MULTI-VALUED FIELD: *See* Repeatable Field

NANO-: *See* Kilo-

NATIVE CODE: *See* Machine Language

NON-PROCEDURAL LANGUAGE: *See* Programming Language

NON-REDUNDANCY: A basic principle of data base organization: the same information must not appear in more than one place within the set of data files. This principle does not apply to working files, which are always redundant with respect to the data files. *See also* Data Base; Data File; Redundancy; Working File

NORMAL FORM: *See* Table

NORMALIZE: (1) To edit a data value for processing or simply for manual look-up, as when ''Dec. 15, 1936'' is normalized to ''12/15/36'' or further normalized to ''19361215''; or ''George Washington'' may be normalized to ''WASHINGTON GEORGE''; (2) In relational data base theory, the reorganization of data into relational tables of progressively higher ''normal form'' as discussed under ''Flat Files and Relations'' in Chapter 4. *See also* Flat File; Relational; Table (usage 1); Value

NULL FIELD: A field with no content, not even unused space. Its length is zero or only one or two bytes to hold the position and indicate the null state of the field. *See also* Field

OBJECT PROGRAM: A program or module in machine language, hence ready for execution without further processing. *See also* Machine Language; Program

OCR: *See* Optical Character Recognition

ON-LINE: Automatic processing with user participation, as opposed to Batch Processing. *See also* Batch Processing.

OPERATING SYSTEM: A master program that controls the operation of all other software, initiating execution, allocating resources (e.g. space in main storage or access to external storage), performing routine chores, and resuming control of the system when other software terminates or raises an error condition. The operating system is the first software loaded into main storage when a computer is turned on and, because of its supervisory role, at least part of the operating system must be present at all times. *See also* Generation (usage 1); Program; Software; Storage

OPTICAL CHARACTER RECOGNITION: Discussed under ''Content Recognition and OCR'' in Chapter 10. No connection with optical disks or optical fibers.

OPTICAL DISK: A data storage medium consisting of a reflective metal and plastic disk bearing concentric tracks of minute pits ''burned in'' by pulses of laser light to represent the stored signal. In reading, a less intense beam of laser light, reflected from the disk surface, flickers as the pits interrupt reflection; the pattern of interruptions conveys the signal to a light sensor. The data on an optical disk cannot be changed; however, its capacity is hundreds of times greater than that of magnetic disks. A small optical disk is called a compact disc when used for sound, a CD-ROM (compact disc read-only-memory) when used for digital data. *See also* ROM; Storage (usages 2b, 2c); Track

OPTION: A user's choice among several possible courses of action offered by software. Simple options are often selected by number from a menu. More complex options may be expressed in the "parameter list" portion of a command. A default option is one taken automatically if the user makes no choice. *See also* Command; Menu; Software

OPTIONAL FIELD: A field that may be omitted from any given record or a repeatable field where the number of occurrences may be zero. In a fixed format file, a field that may be left empty. *See also* Field; Fixed Format; Record; Repeatable Field.

OUTLINE IMAGE: An image represented only by raster points at its angles, straight lines connecting certain of the points, and sometimes color values associated with enclosed areas; as opposed to an image represented by an array of pixels. *See also* Pixel; Raster Point

OUTPUT: (1) The issuance of an intelligible signal from an electronic data processing system for display on screen, paper, or microfilm, either to control some external device, for transfer to another system via tape, moveable disk, or communications line, or for storage and eventual reentry to the same system; (2) That which is output (usage 1). *See also* Input

PAGE: *See* Physical Record

PAINT: In computer graphics, the process of filling an enclosed area with color, pattern, or shade or tracing a line more than one pixel in width. *See also* Pixel

PARAGRAPH: In connection with structured text files, a named field reserved for discussion of some limited aspect of the subject, such as "history" or "description," is sometimes called a paragraph. Technically, this is a text field. *See also* Field; Text; Text Field

PARAMETER LIST: *See* Command

PASSWORD: A string of characters known to the operating system and to one or more particular users, supposedly secret from all others. A user seeking access to restricted programs or data must enter a valid password. Upon recognition, the operating system looks up the user's account to determine what access is authorized. At every installation there is at least one password that gives access to everything, including the file of passwords. *See also* Character; File; Operating System

PERMUTATION: A valuable option in procedures that produce a sorted list of items. If items are sorted upon repeatable or multi-valued fields, permutation replicates the item, listing one copy under each value. Thus an object with "material: ivory" and also "material: wood" would appear under "ivory" and again under "wood" in a list sorted by material. *See also* Option; Repeatable Field; Sort; Value

PHYSICAL FILE: The physical storage area reserved to contain a file. It has a unique name, a location, a size or "length," and one or more equal divisions called physical records. The physical file is not a body of data but only a container for data. The content, if any, constitutes a logical file or a part of one. *See also* File; Logical File; Physical Record; Record

PHYSICAL RECORD: One of the uniform divisions of a physical file. Physical records are to a physical file what pages are to a volume, and they are sometimes called "pages." A physical record may hold one logical record, several, part of one, or it may be empty, just as a page may hold one or more paragraphs or may be blank. In many cases the physical records are numbered, as pages are, to facilitate the search for data. *See also* Logical Record, Physical File

PICO-: *See* Kilo-

PIXEL: Short for *picture element.* One of an array of small rectangles comprising the digital representation of an image as opposed to an outline image. Each pixel is recorded with from one to three numerical values, the minimal representation being a single bit standing for black or white. *See also* Array; Bit; Digital; Outline Image

POINTER: (1) Any coded reference within a record that enables the system to find and retrieve other records from the same file or elsewhere in the data base. In non-relational systems a pointer is usually a physical storage location: file name and physical record number. In relational systems it is usually a key field, e.g. an accession or object number. (2) The data type of a field containing a physical location pointer, primarily for system use. *See also* Data Base; Data Type; File; Key; Linked Record; Physical Record; Record

POSTPROCESSING: Data processing between retrieval and output. Examples include record counting and derivation of statistics, inserting diacritical marks and punctuation, conversion of units of measure, decoding, vocabulary substitution—from English to French terminology, and preparation of visual displays such as charts and maps. A step in the report generation process. *See also* Coding (usage 2); Output; Report; Retrieval

PRECISION: A measure of the size or length of a numeric field. For example, a byte has a precision of six to nine bits, usually eight. An eight-bit byte may also be said to have a precision of 256 because it contains exactly 256 possible combinations of bits with values of zero or one. A decimal numeric field of three characters has a precision of three digits or 1,000 values ("000" through "999"). Precision is independent of scale. This means that a three-digit field has the same precision regardless of whether it represents values from ".000" through ".999" or "000" through "999." *See also* Bit; Byte; Character; Data Field; Data Type; Value

PREPROCESSING: Data checking and other processing that occurs between input and transmission to external storage. Examples include spelling verification, checking against authority files, range checking, syntax checking, conversion from English to metric units of measure, and encoding. *See also* Authority List; Coding (usage 2); Input; Storage

PRIMARY KEY: *See* Key

PRIMITIVE DATA: Data as input, not computed or selected from other data; as opposed to derived data. *See also* Data File; Derived Data; Input

PROCEDURE: (1) A self-contained program within a program, or a "module" invoked by another program; (2) A program of programs, also called a stored procedure. A list of programs or "steps" to be executed in order whenever the procedure is invoked. Many procedures also list the files to be processed at each step and specify conditions for skipping or repeating some programs. A labor-saving device wherever the same programs are often used in the same order; (3) A file containing a stored procedure. *See also* File; Module; Program

PROCEDURE FILE: A file containing a procedure (usage 2) or a sequence of commands. The name of a procedure file may be used as a single high-level command to initiate execution of all the component commands in order. *See also* Command; File; Procedure

PROCESSING UNIT: The heart of a computer and the only site where data are actually altered or processed, as when a 2 and a 3 go in for addition and a 5 comes out. Each processing unit consists of a set of instructions (usage 2), usually between 100 and 300. Because no two processing units include exactly the same instruction set, every machine language program is compatible with only one processing unit. Some microcomputers contain more than one, to accommodate a wider range of software. Some mainframe computers have two or four processing units working in parallel to speed processing. So-called supercomputers may have many processing units. Most computer companies do not manufacture their own processing units but purchase them from a much smaller number of microchip manufacturers. Thus many computers have identical processing units and use the same off-the-shelf software. Processing units are customarily "named" with numbers such as the Intel 80286 and 80386 and the Motorola 68020. *See also* Instruction; Machine Language; Microchip; Software

PROCESSOR: (1) Any program or system of programs, e.g. a "word processor"; (2) A processing unit. *See also* Program; Processing Unit

PROCESSOR CHIP: A microchip containing a processing unit. *See also* Microchip; Processing Unit

PROGRAM: A series of specifications for data processing by computer, prepared according to the rules of some programming language, and, ultimately, stored in a program file for processing. Programs contain lesser programs or "procedures," invoke other programs called "modules" or procedures, and are contained within greater programs or "systems." Hence just what a program is depends upon one's point of view at the moment. Only "object" programs in machine language can be executed directly. All others must first be assembled, interpreted, or compiled to produce an executable object program. *See also* Machine Language; Module; Object Program; Procedure; Program File; Programming Language

PROGRAM FILE: A file containing a computer program or a system of programs or program modules, not a user's data. File contents may be a "source program" written in some programming language or an object program "compiled" into absolute machine language or "native code" suitable for execution by a computer. Some programs are also stored in intermediate stages, i.e., partially compiled. A source program is a text file in programming language. *See also* File; Machine Language; Module; Object Program; Program; Programming Language; Source Program; Text File

PROGRAM GENERATOR: *See* Programming Language

PROGRAMMING LANGUAGE: A set of rules for controlling a computer by means of input. Hundreds if not thousands of programming languages have been developed. They are classified according to "level," with machine language representing the lowest level, though some languages now fall between the traditional levels:

Machine language programs consist of a series of instructions, each of which is a short series of binary numbers.

Assembly language is like machine language except that each instruction is represented by a combination of alphabetic and numeric symbols that can be understood by a person, but must be translated or "assembled" into machine language by a program called an assembler.

Macro languages are like assembly languages except that they include some "macro statements" that specify a series of instructions, not just one.

High-level languages such as FORTRAN, COBOL *et al.,* use statements that express the user's intent without reference to specific machine instructions. Hence they may be compatible with more than one processor, i.e., the code is transportable. Such programs are either "compiled" into machine language by a compiler program or "interpreted," which means that each statement is translated individually into machine language as it is executed and no complete machine language version is produced. BASIC is usually interpreted, though it can also be compiled.

Non-procedural or *fourth-generation languages,* in theory, specify only the required result, leaving the actual sequence of events to a species of fourth-generation software called a program generator. No current language is totally non-procedural, but many are so classified. RPG was the original product of this type. *See also* Binary Number; Instruction; Machine Language; Processor; Program; Software

PROMPT: In interactive data processing, the system's signal to the user that it is ready and waiting for input, either data or instruction. The usual prompt is the appearance of a symbol on the screen, often merely a period or a right angle bracket (⟩), sometimes accompanied by an audible tone or "beep." A prompt may also include a menu of options. *See also* Input (usage 2); Menu; Option

QUERY: A formal statement specifying what logical records are to be retrieved in a file search. A query must be formulated according to the query language of the system in use. Query statements are classified as follows: (1) A *simple query* states a single search criterion, e.g. FIND (object='doll'); (2) a *range query* states a numerical or, less often, an alphabetical range of values to be retrieved, e.g. (year=1837–1901); and (3) a *boolean query* combines two or more simple or range queries in a single statement, using parentheses as in algebra plus the three boolean operators AND, OR, and NOT, e.g. FIND (((object='doll') OR (object='dollhouse')) AND (year=1837–1910) AND (NOT (condition='fragment') OR (condition='poor'))). *See also* File; Language; Logical Record; Retrieve

RAM (R*andom* A*ccess* M*emory*): *See* Storage

RANGE QUERY: *See* Query

RASTER POINT: A single point, designated by a pair of numbers, in an outline image; comparable to the intersection of lines of latitude and longitude on a map. Similar to a pixel except that a pixel is a small rectangular area with some color value, while a raster is only a point, to and from which lines may be drawn. *See also* Outline Image; Pixel

READ: To copy data from any external or archival storage or from an input device into internal storage. Reading is an I/O operation, meaning that no matter what the storage medium, some mechanical motion is necessary. Copying does not remove data from an external medium. Hence data are never written back after reading unless they have been changed permanently. *See also* Input; I/O; Storage; Write

REAL NUMBER: *See* Floating Point Number

RECORD: One of the logical or physical units (items, paragraphs, or rows) that comprise a file. *See also* File; Item File; Logical File; Logical Record; Paragraph; Physical File; Physical Record; Row

RECORD FORMAT: For any logical file except a pure text file, a permanently stored data definition describing all component fields and domains of each logical record. *See also* Data Definition; Data Field; Domain; Logical File; Logical Record; Text File

REDUNDANCY: Occurrence of the same information more than once in a data base. The principle of non-redundancy forbids redundancy within the set of data files, but not among working files. *See also* Data Base; Data File; Non-Redundancy; Working File

RELATE: *See* Join

RELATION: In a relational data base, a flat file in which the records are called rows or tuples and the corresponding fields of each row constitute columns. Discussed under "Flat Files and Relations" in Chapter 4. *See also* Column; Field; Flat file; Record; Row; Table

RELATIONAL: Describes a data base management system supporting a data base of relations and certain data base operations, of which the most important for museum work is "joining." A full definition is beyond the scope of this book. Not all software advertised as relational really is. *See also* Data Base; Data Base Management System; Join; Relation

REPEATABLE FIELD: A field that may appear more than once in any given record. In some systems each repeatable field has a maximum allowable number of repetitions, but repetition may also be unlimited. In catalog records, repeatable fields are useful for lists of such multiple categories as former owners or materials. They are also called multi-valued fields. Repeatable fields are forbidden in flat files and tables. *See also* Field; Flat File; Record; Table; Value

REPEATING GROUP: A repeatable composite field. *See also* Composite Field; Repeatable Field

REPORT: Organized, useful output of a data processing system, usually in "hard copy" but sometimes also on a screen. *See also* Hard Copy; Output

REPORT GENERATOR: A program, usually included in a DBMS, that edits and combines data with labels, punctuation, spacing, pagination, etc., and sometimes creates graphs or diagrams for output via display or printing. *See also* Data Base Management System; Forms Design; Output; Program

RESOLUTION: In optics, a measure of the closeness of points that can be distinguished separately, hence, in a digitized image, the number of separate raster points or pixels; the fineness of detail. *See also* Digital; Pixel; Raster Point

RETRIEVE: To read data from external storage. *See also* Read; Storage

ROM (R*ead* O*nly* M*emory*): *See* Storage

ROW: *See* Table

SCANNER: A hardware device for digitizing a flat image such as a printed page. Output is an array of pixel descriptions. *See also* Array; Hardware; Pixel; Scanning (usage 2)

SCANNING: (1) Sequential searching (usage 2); (2) The process of digitizing an image by sensing light and dark, point-by-point across the image and line-by-line from top to bottom; the first step in optical character recognition (OCR). *See also* Optical Character Recognition; Sequential Searching

SCIENTIFIC NOTATION: *See* Floating-Point Number

SECTOR: A unit of physical storage on a "sectored disk" amounting to a fraction of one track, e.g. 1/13th. Many small systems use fixed, 512-byte sectors. With some DBMSs the length of a sector is also the maximum length of a physical record. Sometimes a sector is also the length of a block of data retrieved on one file access. *See also* Block; Byte; File Access; Physical Record; Storage; Track

SELECTION: (1) In relational data base processing, copying rows from a *relation* for output or to build a new relation (or working file); (2) Copying specific records, fields, or subfields for use in computation, report generation, or incorporation into another file. *See also* Field; File; Record; Relation; Report; Row; Subfield; Working File

SEQUENTIAL SEARCHING, SEQUENTIAL ACCESS: (1) Reading a file's physical records in the order in which they are stored. It requires that logical records correspond to physical records at least to the extent that all parts of a logical record are in contiguous physical records. When files are read from tape this is the only possible search strategy; (2) Scanning a text file or free text field, from the beginning, for a specified character or sequence of characters. *See also* Character; File; Free Text; Logical Record; Physical Record; Read; Text File

SOFTWARE: (1) Originally, the intangible components of an electronic data processing system; distinguished from tangible components called "hardware"; (2) Control programs or "system software" such as operating systems, language compilers, and data base management systems, supplied by a computer or software vendor—excludes user-written programs and procedures. Also data files supplied by a vendor to support system software, i.e., a spelling file supplied with word-processing software. *See also* Data Base Management System; Data File; Hardware; Operating System; Procedure; Program; Programming Language

SORT, SORTING: Rearranging the logical records of a file in order according to the values of specified key fields, i.e., by "artist" or "object number." A file may be sorted "in place," so that the sorted file replaces the original in the same location and retains the same name; otherwise, the original may be left as it was and a new file created with the same logical records in a different order. A sort/merge operation sorts two or more files and merges the combined contents to form a new file. *See also* Field; File; Key; Logical Record; Merge; Sort Field; Value

SORT FIELD: Any data field used for sorting or merging records alphabetically or numerically, but especially a field normalized for this purpose. Some sort fields serve no other purpose since they duplicate information to be found elsewhere in the record. Some are generated automatically from un-normalized data. Normalized sort fields may also serve as key fields for indexing records. *See also* Field; Index; Key; Merging; Normalize (usage 1); Record; Sort

SOURCE FILE: (1) Any file from which data are read by a computer program, especially a source program to be compiled, a data file to be sorted, or a file to be combined with another; (2) In a museum, a file relating to the sources of objects in the collection, e.g. a donor file, vendor file, or a file of former owners. *See also* Donor File; File; Programming Language; Sort; Source Program; Vendor File

SOURCE PROGRAM: A computer program not in machine language, hence not executable without assembly, compilation, or interpretation. *See also* Machine Language; Programming Language

STATEMENT: *See* Programming Language

STORAGE: A repository of digitized data. All data within a computer system are stored. Even when data are in motion a copy remains stored at the point of origin. There are many kinds of storage:

1. *Internal storage* is directly accessible to the processing unit without any mechanical action and is primarily in the form of *volatile storage,* which is an array of transistors used as electronic switches, each in the open or closed state. Since a small current is necessary to hold the switches in position, all data in volatile storage vanishes if the system is turned off or the power interrupted. Three main types of volatile storage are

> a. *Main storage,* also called RAM in microcomputers or ''core,'' holds data and programs currently in use
>
> b. *Registers* are word-sized units of ultra-fast storage for data needed immediately or constantly by the processing unit
>
> c. *Cache storage* holds records retrieved from external storage in anticipation of probable need. The records must be moved to main storage before they can be processed.

ROM is a nonvolatile form of internal storage that holds built-in system software that cannot be changed without replacing the physical ROM storage unit. Used primarily but not exclusively in microcomputers.

2. *External storage* is permanent, but data are inaccessible to the processor until ''retrieved'' or copied into internal storage, which always entails some mechanical action. There are several forms of external storage:

> a. *Magnetic tapes* and *disks* are erasable storage—the system can write data upon them as well as retrieve data from them
>
> b. Most *optical disks* are a form of external ROM—the system can read data from them but not write upon them.
>
> c. *WORM* (*w*rite *o*nce, *r*ead *m*any) storage uses optical disks that are initially blank. The system can write data upon them but cannot change or delete what has been written.

3. *Archival storage* is off-line, meaning that human intervention is necessary to mount a tape or disk or to feed punched cards or tape through an input device before archived data can reach external or internal storage. Archived data may reenter the system that stored it or any other system compatible with

the first. The media are moveable and capacity unlimited. *See also* Input; Optical Disk; Processing Unit; Read; Record; Software; Word; Write

STRING: A data field consisting of an ordered series of bytes or bits such as the character string "character string" or the bit string "0010000." A portion of the "character string" string, such as "racter," is called a substring, while a portion including the beginning of the string, such as "char," is a leading substring. A trailing substring is one that includes the end of the string. *See also* Bit; Byte; Character; Data Field; Programming Language

SUBFIELD: A subdivision of a field or a "field within a field," e.g. an object location field with subfields for: building/room/storage unit/shelf number/date verified. It is possible, though unusual, to have sub-subfields and so on to any degree of division within divisions. *See also* Field

SUBSTRING: *See* String

SYNTAX: Rules of data representation applicable to a data field or domain. For example, "Dec. 15, 1936," "15 December 36," "36-12-15," and "19361215" all convey the same information, but differ in syntax. *See also* Data Field; Domain; Normalize (usage 1)

TABLE: (1) In connection with relational data bases, a flat file in which the records are called rows and the fields columns. Relational tables are classified as being in first, second, third, or fourth "normal" form, each successive form subject to additional constraints as to logical relationships or dependencies among its columns; (2) In connection with report generation, representation of data in a tabular format. *See also* Field; Flat File; Record; Relation; Report

TABLE OF CONTENTS: A list of physical file locations, usually for one magnetic tape or magnetic or optical disk, maintained and used automatically by the operating system in conjunction with directory file. Generally invisible to the user. *See also* Directory; Operating System; Optical Disk; Physical File

TEMPLATE: An empty logical record form displayed on a screen as a guide for data input. Usually the template lists component fields and subfields in order, each followed by blank spaces corresponding to maximum field length. *See also* Field; Input; Length; Logical Record; Subfield

TERA-: *See* Kilo-

TEXT: *See* Character (usage 4)

TEXT DATA BASE: A data base in which all data files are structured text files and indexing is by paragraph and word position. An example is IBM's DBMS called STAIRS (TM). *See also* Data Base; Data File; Paragraph; Text File; Word Indexing

TEXT FILE: A file containing only the keyboard characters, i.e., numbers, letters, punctuation, and special characters, plus such control characters as back space, tab, space, carriage return, page feed, and end of file. *See also* Character; File

THESAURUS: A list of terms, similar to an authority list, in which each term may be associated with one or more of the following sublists: preferred term; synonymous term(s); narrower term(s); and broader term(s). *See also* Authority List

TRACING: In history museum cataloging, a list of an object's historical associations with persons and events; not to be confused with tracking

TRACK: (1) On magnetic tape, one of seven or nine parallel lines representing stored bits. On tape, each byte is represented by a transverse row of bits, one in each track across the tape. Each byte contains one extra bit used by the system. Hence a seven-track tape corresponds to a six-bit byte, a nine-track tape to an eight-bit byte; (2) On magnetic and optical disks, one of many hundreds of concentric circular lines representing stored bits. Disk tracks are much more closely spaced than tracks on tape and the bits of each byte are in series, one after another, in the same track. Only one track can be read at a time. On disks, every track has the same number of bits, bytes, and sectors (if the disk is sectored). Therefore, each bit is longer on an outer track than on an inner track and the same number of bits are read per disk revolution regardless of the track's position. *See also* Bit; Byte; Optical Disk; Read; Sector; Storage

TRACKING: In collection documentation, a cumulative historical list of an object's physical movements and locations

TREE, TREE STRUCTURE: An organization of linked entities or "nodes" such that each except the first or "root" is linked to one "higher" node and all except the "leaves" are linked to two or more deeper nodes. The nodes of a tree may represent fields, logical records, "pages" of a tree-structured index, files, data bases, machines, or computer systems—the term "tree" refers only to the pattern of links. Normally trees are diagrammed upside-down (see Figure 12). *See also* Data Base; Field; File; Index; Linked Record; Logical Record; Network

TRUTH FILE:—See Data File

UNDELETE: *See* Delete

UNORDERED FIELD: A field that is identified not by its position (first, second, etc.) within the record but by a label attached to the field. Useful in files such as museum catalogs where a large number of fields are defined, but only a small number may be applicable to any given object. *See also* Field; File; Record

UPDATE: To change the content of an existing file by appending, inserting, or deleting logical records, fields, subfields or by changing the values of fields or subfields. *See also* Append; Delete; Field; File; Insert; Logical Record; Subfield; Value

VALUE: The meaningful content of a logical field of any data type, not necessarily numerical. *See also* Data Field, Data Type; Logical Record

VARIABLE LENGTH: In connection with fields of the text or character data type, a field in which a varying number of characters (usually up to some maximum length) may be entered. An empty field of variable length has an actual length of zero and is called a null field. In contrast, a nonvariable field that is not filled to capacity contains blanks at the end to fill all available space. *See also* Character; Data Type; Field; Length; Null; Text

VDT: *See* Cathode Ray Tube

VENDOR FILE: (1) A list of sources of hardware, software, supplies, or any commodity, usually including a summary of what the vendor sells plus other information and comments; (2) In a museum, a file of those from whom objects have been purchased for the collection, each vendor being associated with a list of items sold to the museum.

VOXEL: Short for "volume element." A voxel is the three-dimensional analog of the pixel and represents a small cubic volume of a three-dimensional image. It is recorded as a series of numeric values representing color, index of reflection, transparency, etc., the minimal representation being a single bit standing for solid or empty. *See also* Bit

WIDTH: *See* Length

WINDOW: An area of a terminal display screen, usually rectangular in shape, devoted to one display relating to one operation. When the screen is divided into several window areas, each is comparable to a separate display screen. Window areas frequently overlap so that one appears to lie behind another, partially obscured, but the user has the option of either removing one window to uncover another or changing the apparent stacking of windows (as if they were papers on a desk) so that any desired window appears to be "on top."

WORD: In computer systems, a group of bits that move in parallel and may be processed together. Depending upon data type, a word may represent one or more bytes or characters or one or more numbers, such as short or long integers. Since word size or "path width" is a factor in processing power and speed, computers are classified according to word size:

• 4-bit machines	The first microcomputers, now obsolete
• 8-bit machines	The microcomputer standard c. 1980
• 16-bit machines	Modern microcomputers
• 32-bit machines	Microcomputers and minicomputers
• 36-bit machines	Minicomputers
• 40–64-bit machines	Mainframes

See also Bit; Byte; Character; Data Type; Integer

WORD INDEXING: Indexing based upon the occurrence of words and phrases in free text or text files, as opposed to indexing upon entire field contents. A word index may locate subjects not only to the logical record, but by paragraph, sentence, and word count within records, especially in a text data base. ''Word'' indexing has nothing to do with the technical definition of ''word'' in this glossary. *See also* False Drop; Field Indexing; Free Text; Index; Logical Record; Paragraph; Text Data Base

WORKING FILE: Goal-oriented, temporary files of data derived from permanent data files by copying, selection, recombination, and computation. Working files are by nature redundant and may, in time, become inconsistent with current data files. *See also* Data File; Derived Data; Redundancy; Selection

WRITE: To copy data from internal storage to external or archival storage or to an output device. Writing is an I/O operation, meaning that some mechanical motion is necessary in all cases except output to a screen or communication line. Writing does not remove the data from internal storage. *See also* I/O; Output; Storage

Selected Bibliography

The references that follow are not intended to be exhaustive. Rather, they are an attempt to provide (1) some sense of historical perspective in what has been a rather esoteric field of specialization, (2) recent publications that we believe will be of greatest value to a wide variety of readers, and (3) bibliographic information on textual citations. Probably the most complete recent bibliography on this subject is included in Light, Roberts, and Stewart 1986.

INFORMATION RESOURCES

Clearinghouse Project, Office of Library Systems, The Thomas J. Watson Library, The Metropolitan Museum of Art, Fifth Avenue at 82nd Street, New York, NY 10028, USA. Maintains, as a resource for other museums, all publicly available data on currently active museum computer projects.

ICOM Documentation Centre, 1, rue Moillis, 75015 Paris, France. ICOM is the International Council of Museums. With limited staff, the Documentation Centre has for many years attempted to keep a current inventory of museum computer projects worldwide, and has served as the sponsoring agency for CIDOC (the International Council of Museums Documentation Committee). In addition, they have occasionally sponsored international conferences and published bibliographies on the subject in book form, although financial and other difficulties have resulted in the bibliographies appearing much later than would be desired.

Museum Computer Network, Inc., P.O. Box 111, East Winthrop, ME 04343, USA. A network in the sense of museums and individuals joining together to share information about standards, principles, technology, and experience, rather than establishing a physical communications network. Now a membership organization. Holds annual meetings and publishes *Spectra* (see next section on periodicals).

Museum Services International, 1716 17th Street, NW, Washington, DC 20009, USA. Telephone: 202-462-2380. A nonprofit company that provides a variety of services to museums, including consulting on computer systems.

Museum Documentation Association, Building "O," 347 Cherry Hinton Road, Cambridge CB1 4DH, England. A network in some ways similar to the Museum Computer Network in the United States, but far more structured. Has published an elaborate system of recording standards developed by national committees of subject specialists, designed for both manual and automated information procession. The MDA also supports a museum software package called GOS, which is available either to museums with their own computers or for input, storage, and processing of data by the MDA, as a service bureau.

Visual Resources Association, Vocabulary Coordination Group, The Getty Art History Information Program, 401 Wilshire Boulevard, Suite 1100, Santa Monica, California 90401, USA. A reference source for people concerned with visual materials, with information on technological developments, professional guides for the field, workshops, programs, meetings, and contacts with professionals worldwide.

PERIODICALS

Computers and the Humanities, Paradigm Press, Inc., P.O. Box 1057, Osprey, FL 34229, USA. Ed. Glyn Holmes. Quarterly, v. 1, 1967, to date.

Curator, American Museum of Natural History, Central Park West at 79th Street, New York, NY 10024, USA. Ed. Thomas D. Nicholson. Quarterly, v. 1, 1958, to date.

——— 30, no. 2 (Jun 1987): A special issue devoted to papers of the Museum Computer Network Annual Conference, Mexico City, Oct. 24–25, 1985. Includes papers by Joan Bacharach, Bonnie Burnham, Robert G. Chenhall, Colin Eades, Jon Gartenberg, Ronald Kley, Sandra Parker, and Susan Wheeler.

International Journal of Museum Management and Curatorship, Butterworth Scientific, Ltd., P.O. Box 63, Westbury House, Bury Street, Guildford GU2 5BH, England. Eds. Peter Cannon-Brookes and Caroline Cannon-Brookes. Quarterly, v. 1, 1982, to date.

MDA Information, Museum Documentation Association, Building "O," 347 Cherry Hinton Road, Cambridge CB1 4DH, England. Ed. D. Andrew Roberts. Quarterly, v. 1, 1977, to date.

——— 6, no. 4 (Jan 1983): 12 pp. devoted to problems of cataloging "backlog" collections and regaining control of museum inventories.

Museum, UNESCO, 7 place de Fontenoy, 75700 Paris, France. Ed. Marie-Josee Thiel. Quarterly, v. 1, 1950, to date.

——— 25, no. 1 (1971); 30, nos. 3/4 (1978): Issues devoted to museums and computers.

Museum News, American Association of Museums, 1225 Eye Street, NW, Suite 200, Washington, DC 20005, USA.. Ed. Ligeia Z. Fontaine. Bi-monthly, v. 1, 1923, to date.

——— 51, no. 8 (Apr 1973): Issue devoted to museums and computers.

Museums Journal, Museums Association, 34 Bloomsbury Way, London WC1A 2SF, England, v. 1, 1901, to date.

——— 82, no. 2 (Sep 1982): A special issue devoted to computer applications in museums. Contains papers by John Burnett and David Wright, Peter David and Judith Hebron, Richard B. Light, Martin Norgate, Elizabeth Orna, Michael J. Seaborne and Steven Neufeld, and Brian Abell Seddon.

Slide Buyer's Guide, Visual Resources Association, c/o Department of Art, James Madison University, Harrisonburg, VA 22807, USA. Ed. Norine Cashman. Order directly from Libraries Unlimited, Inc. Attn. Dept. 75, P.O. Box 263, Littleton, CO 80160-0263, USA.

Spectra, Museum Computer Network, Inc., P.O. Box 111, East Winthrop, ME 04343, USA. Ed. Julia A. Hunter, Quarterly, v. 1, no. 1, Winter 1974, to date.

Visual Resources: An International Journal of Documentation, Visual Resources Association, c/o Department of Art, James Madison University, Harrisonburg, VA 22807, USA. For subscription information, write to Gordon and Breach Science Publishers, Inc. P.O. Box 786 Cooper Station, New York, NY 10276, USA; in UK, P.O. Box 297, London, WC2E 9PX.

ARTICLES AND BOOKS

Abu-Mostfa, Yaser S. and Demetri Psaltis. 1987. Optical neural computers. *Scientific American* 256, no. 3 (Mar 1987): 88–95.

ASM (American Society of Mammalogists, Committee on Information Retrieval). 1984. *Survey Report on Computerized Information Retrieval in Mammal Collections of North America.* Pittsburgh: ASM.

Art Museum Association of America. 1982. *Technology in Museum Environments.* San Francisco: Art Museum Association of America.

———. 1984. *A Reference List of In-House Computer Use in Art Museums.* San Francisco: Art Museum Association of America.

Aubert, Michel and Dominique Piot. 1986. Documenting French cultural property. (In) Light, Roberts, and Stewart 1986: 233–40.

Banta, Mellissa, Germaine Juneau and Lea S. McChesney. 1986. Saving the sacrificed: exhibition collaboration and computerization. *Museum News* 65, no. 1 (Oct/Nov 1986): 48–53.

Bearman, David. 1987. *Technical Report on Functional Requirements for Collections Management.* Pittsburgh: Archives & Museum Informatics.

Bergengren, Gören and Heidi Henriksson. 1986. Nordiska museet. (In) Light, Roberts, and Stewart 1986: 267–74.

Bisongi, Fabio. 1978. *The Catalogue of Italian Art with Iconographical Analysis Realized with the Use of the Computer.* Siena: Universita di Siena.

Bourelly, Louis and Eugene Chouraqui. 1978. *Le System Documentaire SATIN 1.* 2 vols. Marseille: Centre National de la Recherche Scientifique.

Burnham, Bonnie. 1978. *Art Theft and the Role of an Art Theft Archive.* New York: International Foundation for Art Research, Inc. (Includes an appendix with an outline of a computerized system for the identification of found objects—see Vance 1984).

Cameron, Duncan F. 1970. Museums, systems and computers. *Museum* 23, no. 1 (1970/71): 11–17.

Castillo Tejero, Noemi. 1972. Diccionario de terminos basicos para describir y catalogar las colecciones arqueologicas del Museo Nacional de Antropologia de Mexico. *Antropologia Matematica* 21. Mexico City: Museo Nacional de Antropologia.

———. 1986. The National Museum of Anthropology, Mexico. (In) Light, Roberts, and Stewart 1986: 103–09.

Cato, Paisley S. and Joseph Folse. 1985. A microcomputer/mainframe hybrid system for computerizing specimen data. *Curator* 28, no. 2 (Jun 1985): 105–16.

Chenhall, Robert G. 1968. The analysis of museum systems. (In) *Computers and Their Potential Applications in Museums,* pp. 59–79.

———. 1971. The archaeological data bank: a progress report. *Computers and the Humanities* 5, no. 3 (Jan 1971): 159–69.

———. 1973. Sharing the wealth. *Museum News* 51, no. 8 (Apr 1973): 21–23.

———. 1975. *Museum Cataloging in the Computer Age.* Nashville: American Association for State and Local History.

———. 1976. Museum Information Networks. *Museum Data Bank Research Report* no. 5.

———. 1977. The onomastic octopus. *Museum Data Bank Research Report* no. 10.

———. 1978. *Nomenclature for Museum Cataloging: A System for Classifying Man-Made Objects.* Nashville: American Association for State and Local History. (Now out of print; a revised edition is scheduled for publication in April 1988).

Choate, Jerry R., Robert C. Dowler and Mark D. Engstrom. 1977. *Standards for Documentation and Data Capture for Mammals Housed in the Museum of the High Plains.* Fort Hayes, KS: Museum of the High Plains.

Cok, Mary V. 1981. *All in Order: Information Systems for the Arts.* New York: Publication Center for Cultural Resources, Inc. (625 Broadway, New York, NY 10012).

Computers and Cultural Materials Conference. 1982. A conference held at the North Carolina Museum of History, 1982. Raleigh: North Carolina Museum of History.

Computers and Their Potential Applications in Museums. 1968. A conference sponsored by the Metropolitan Museum of Art, April 15–17, 1968. New York: Arno Press.

Cuisenier, Jean. 1970. Feasibility of using a data-processing system in the Musee des Arts et Traditions Populaires, Paris. *Museum* 23, no. 1 (1970/71): 27–36.

Dauterman, Carl. C. 1968. Sevres incised marks and the computer. (In) *Computers and Their Potential Applications in Museums* pp. 177–94.

De Borhegyi, Stephen and Alice Marriott. 1958. Proposals for a standardized museum accessioning and classification system. *Curator* 1, no. 2 (Spring 1958): 77–86.

Delroy, Stephen H. 1987. Data standards. *Spectra* 14, no. 3 (Fall 1987): 3–6.

Dixon, Rob. 1983. A modern computer cataloguing and administration system for museums. *International Journal of Museum Management and Curatorship* 2, no. 4 (Dec 1983): 335–46.

Ellin, Everett. 1968. An international survey of museum computer activity. *Computers and the Humanities* 3, no. 2 (Nov 1968): 65–86.

———. 1969. Museums and the computer: an appraisal of new potentials. *Computers and the Humanities* v. 4, no. 1 (Sep 1969): 25–30.

Ferrari, Oreste. 1986. Documentation of the Italian cultural heritage. (In) Light, Roberts, and Stewart 1986: 243–53.

Fineberg, Ellen. 1983. Hi-tech history. *History News* 38, no. 9: 23–26. (Deals with videodisc applications).

Folse, L. Joseph and Paisley S. Cato. 1985. Software needs for collection management. *Curator* 28, no. 2 (Jun 1985): 95–104.

Freundlich, A. L. 1966. Museum registration by computer. *Museum News* 44, no. 6 (Feb 1966): 18–20.

Gartenberg, Jon. 1980. *Film Cataloguing Manual: A Computer System*. New York: Museum of Modern Art.

Gautier, T. Gary. 1986. National Museum of Natural History, Smithsonian Institution. (In) Light, Roberts, and Stewart 1986: 48–54.

Golden, Julia and Leslie F. Marcus. 1982. Computer generated catalogues for a fossil invertebrate type collection. *Curator* 25, no. 2 (Jun 1982): 107–19.

Goodwin, Larry and Mary Ellen Conaway. 1984. The micro and the muse. *Museum News,* 62, no. 4 (Apr 1984): 55–63.

Green, Dee F. 1968. A prototype computerized system for archaeological collections. (In) *Computers and Their Potential Applications in Museums,* pp. 39–57.

Gundlach, Rolf. 1968. Zur maschinellen Erschliessung historischer Museumbestaende. *Museumskunde* 3: 135–46.

———. 1970. Grundlagen und Strukturen der Komponentendeskription in der historischen Dokumentation. *Nachrichten fuer Dokumentation* Beiheft 20: 49–69. (Summary in English).

———. 1971. Strukturierung und Beschreibung archaeologischer Objekte in der "Documentation aegyptischer Altertuemer." *Archaeographie* no. 2: 9–77.

Gundlach, Rolf and Carl August Leuckerath. 1969. Nichtnumerische Datenverarbeitung in den historischen Wissenschaften. *Geschichte in Wissenschaft und Unterricht Zeitschrift des Verbandes der Geschichtslehrer Deutschlands* 7: 385–98.

Gundlach, Rolf and Albert Schug. 1970. Zur Bedeutung der Kompatibilitaet in der Museumsdokumentation. *Museumskunde* 2: 79–83.

Heller, Jack. 1974. On logical data organization, card catalogs and the GRIPHOS management information system. *Museum Data Bank Research Report* No. 3.

Hillis, Daniel W. 1987. The connection machine. *Scientific American* 256, no. 6 (Jun 1987): 108–15.

Homulos, Peter S. 1978. The Canadian National Inventory Program. *Museum* 30, no. 3/4: pp. 153–59.

Humphrey, James. 1967. The computer as art cataloguer. *Computers and the Humanities* 1, no. 5 (May 1967): 164–69.

Humphrey, Philip S. and Ann C. Clausen. 1977. *Automated Cataloging for Museum Collections: A Model for Decision and a Guide to Implementation*. Lawrence, KS: Association of Systematic Collections.

Information Problems in Art History. 1982. Proceedings: International seminar held at Oxford, Mar 20–22, 1982. *Art Libraries Journal* 7, no. 2.

Information Retrieval for Museums. 1967. Report of a colloquium held at the City Museum, Sheffield, April 1967. *Museums Journal* 67, no. 2 (Sep 1967): 88–120.

Inmon, William. 1986. Building the best data base. *Computer World Focus* Jul 9, 1986, p. 73.

International Association of Art Critics. 1984. Automation Takes Command: Art History in the Age of Computers. A special issue of *AICARC Bulletin*. Zurich: AICARC.

Kahn, Robert E. 1987. Networks for advanced computing. *Scientific American* 257, no. 4 (Oct 1987): 136–43.

Lewis, Geoffrey. 1965. Obtaining information from museum collections and thoughts on a national museum index. *Museums Journal* 65, no. 1 (Jun 1965): 12–22.

Light, Richard B. and D. Andrew Roberts. 1981. International museum data standards and experiments in data transfer. *MDA Occasional Papers* no. 5.

———. 1984. Microcomputers in museums. *MDA Occasional Papers* no. 7.

Light, Richard B., D. Andrew Roberts and Jennifer D. Stewart, eds. 1986. *Museum Documentation Systems: Developments and Applications*. London: Butterworths.

Lytle, Richard H. 1984. Archival information exchange: a report to the museum community. *Curator* 27, no. 4 (Dec 1984): 265–73.

Manning, Raymond B. 1969. A computer-generated catalog of types: a by-product of data processing in museums. *Curator* 12, no. 2: 134–38.

Markey, Karen. 1983. Computer-assisted construction of a thematic catalog of primary and secondary subject matter. *Visual Resources* 3: 16–49.

———. 1986. *Subject Access to Visual Resources: A Model for the Computer Construction of Thematic Catalogs*. Westport, CT: Greenwood Press, Inc.

McAllister, Don E. 1983. An introduction to minicomputers in museums. *Syllogeus* no. 44: 105–36.

McAllister, Don E. Robert Murphy and John Morrison. 1978. The compleat minicomputer cataloguing computerization and collection inventory. *Curator* 21, no. 1 (Mar 1978): 63–91.

McAllister, Don E. and R. Jon P. Planck. 1981. Capturing fish measurements and counts with calipers and probe interfaced with a computer or pocket calculator. *Canadian Journal of Fisheries and Aquatic Sciences* 38, no. 4: pp. 466ff.

Megna, Ralph J. 1983. Solving big problems with small computers. *Museum News* 62, no. 1 (Oct 1983): 61–66.

Mello, James F. 1977. Computerization of synonymy data from biological systematics. *Museum Data Bank Research Report* no. 9.

Mopberg, Dick and Ira M. Laefsky. 1982. Videodiscs and optical data storage. *Byte* 7, no. 6 (Jun 1982): 142–60.

Murback, David, ed. 1984. *Directory of Computer Use in Plant Record Keeping*. Swarthmore: American Association of Botanical Gardens and Arboreta.

Museum Data Bank Research Reports. 1974–77. The Museum Data Bank Committee published 12 separate research reports before it was disbanded in 1977. The individual reports are listed elsewhere in this bibliography. Remaining stocks are held by the Margaret Woodbury Strong Museum, Rochester, New York.

Museum Documentation Association. 1977. Proposals for the documentation of conservation in museums. *MDA Occasional Papers* no. 1.

———. 1981. *Practical Museum Documentation*. 2nd Ed. Duxford, England: Museum Documentation Association.

Neff, Jeffrey M. and Holly M. Chaffee. 1977. REGIS—a computerized museum registration system. *Curator* 20, no. 1 (Mar 1977): 32–41.

Neufeld, Steven D. 1981. The MDA systems and services: a user's view. *MDA Occasional Papers* no. 6. Duxford, England: Museum Documentation Association.

Oehler, Hansgeorg. 1970. Electronic documentation of a collection of Roman sculpture photographs. *Museum* 23, no. 1 (1970/71): 37–51.

Orchiston, Wayne. 1981. Computers in museums. (In) *Museums and the New Technol-*

ogy: Proceedings of the Annual Conference of the Museums Association of Australia, October 1980, pp. 23–34. Sydney: Museums Association of Australia.

Orna, Elizabeth. 1984. Using a micro to help in thesaurus construction. *MDA Information* 8, no. 3 (Autumn 1984): 66–72.

Orna, Elizabeth and Charles Pettit. 1980. *Information Handling in Museums.* London: Clive Bingley, Ltd.

Painting and Sculpture in the Museum of Modern Art: A Catalog. 1958. New York: Museum of Modern Art.

Peebles, Christopher S. and Patricia Galloway. 1981. Notes from the underground: archaeological data management from excavation to curation. *Curator* 24, no. 4: 225–50.

Peters, James A. and Bruce B. Collette. 1968. The role of time-share computing in museum research. *Curator* 11, no. 1 (Mar 1968): 65–75.

Poggio, Tomaso. 1984. Vision by man and machine. *Scientific American* 250, no. 4 (Apr 1984): 106–17.

Porter, M. F. 1978. Establishing a museum documentation system in the United Kingdom. *Museum* 25, no. 3/4: 169–78.

Porter, M. F., Richard B. Light and D. Andrew Roberts. 1976. A unified approach to the computerization of museum catalogues. *British Library Research and Development Reports,* no. 5338 HC Dec 1976.

Prown, Jules David. 1966. The art historian and the computer: an analysis of Copley's patronage, 1753–1774. *Smithsonian Journal of History* 1, no. 4 (Winter 1966): 17–30.

Pullen, Dennis R. 1985. Inventorying historical collections in the small museum. *Curator* 28, no. 4: 271–85.

Rensberger, John M. and William B. N. Berry. 1967. An automated system for retrieval of museum data. *Curator* 10, no. 4 (Dec 1967): 297–317.

Ricciardelli, Alex F. 1968. Inventorying ethnological collections in museums. (In) *Computers and Their Potential Applications in Museums,* pp. 81–97.

Ricciardelli, Eloise. 1986. The Museum of Modern Art. (In) Light, Roberts, and Stewart 1986: 65–76.

Roberts, D. Andrew. 1985. *Planning the Documentation of Museum Collections.* Cambridge: Museum Documentation Association.

Roberts, D. Andrew and Richard B. Light. 1980. Progress in documentation. *Journal of Documentation* 36, no. 1 (Mar 1980): 42–84.

Rush, Carole E. 1977. An information system for history museums. *Museum Data Bank Research Report* no. 11.

Sarasan, Lenore. 1981. Why museum computer projects fail. *Museum News* 59, no. 4 (Jan/Feb 1981): 40–49.

———. 1986. A system for analyzing museum documentation. (In) Light, Roberts, and Stewart 1986: 89–99.

Sarasan, Lenore and A. M. Neuner. 1983. *Museum Collections and Computers: Report of an Association of Systematics Collections Survey.* Lawrence, KS: Association of Systematics Collections.

Saro, Carlos and Christof Wolters. 1985. *Handbuch Datenerfassung/kleine Museen.* 2 vols. Berlin: Staatliche Museen Preussischer Kulturbesitz, Institut fuer Museumskunde.

Scholtz, Sandra C. 1974. Data structure and computerized museum catalogs. *Museum Data Bank Research Report* no. 2.

———. 1976. A management information system design for a general museum. *Museum Data Bank Research Report* no. 12.

Schug, Albert. 1977. Ein System zur Inventarisation und Dokumentation in den Objektbezogenen Kulturhistorischen Wissenschaften. *Archaeographie* 6: 27–95.

Schuller, Nancy, ed. 1979. *Guide to Management of Visual Resources Collections.* Harrisonburg, VA.: Visual Resources Association.

Schulman, Judith L. 1986. The Detroit Art Registration Information System (DARIS). (In) Light, Roberts, and Stewart 1986: 77–88.

Scott, David W. 1976. The yogi and the registrar. *Museum Data Bank Research Report* no. 7.

Scuola Normale Superiore. 1978. *First International Conference on Automatic Processing of Art Historical Data and Documents, Sep 4–7, 1978.* Pisa: Scuola Normale Superiore.

———. 1984. *Second International Conference on Automatic Processing of Art Historical Data and Documents, Sep 24–27, 1984* (vol. 1—Census of 162 Projects; vol. 2—Papers; vol. 3—Proceedings). Pisa: Scuola Normale Superiore.

Sheffield, University of, Center for English Cultural Tradition and Language, SHIC Working Party. 1983. *Social History and Industrial Classification: A Subject Classification for Museum Collections.* 2 vols. Sheffield: University of Sheffield.

Sher, Jacob. 1986. Museum documentation system and computers: USSR experience. (In) Light, Roberts, and Stewart 1986: 287–92.

Sledge, Jane and Betsy Comstock. 1986. The Canadian Heritage Information Network. (In) Light, Roberts, and Stewart 1986: 7–16.

Small, Joselyn Penny. 1987. *How to Choose a Data Base.* Bronx, NY: American Philological Association.

Squires, Donald F. 1966. Data processing and museum collections: a problem for the present. *Curator* 9, no. 3 (Sep 1966): 216–77.

———. 1968. Collections and the computer. *Bio Science* 18, no. 10 (Oct 1968): 973–74.

Sutherland, Jane and Ginger H. Geyer. 1982. A traditional art museum: modern inventory control. *Perspectives in Computing* 2, no. 3 (Oct 1982): 32–41 (An IBM publication; ref. G324-0007).

Sustik, Joan M. 1981. *Art History Interactive Videodisc Project at the University of Iowa.* Iowa City: Weeg Computing Center, University of Iowa.

Sutton, John F. and Craig C. Black. 1976. Data in systematics collections. *Museum Data Bank Research Report* no. 6.

Swinney, Holman J. 1976. Characteristics of history museum activity and their influence on potential electronic cataloging. *Museum Data Bank Research Report* no. 8.

Vance, David. 1970. Museum data banks. *Information Storage and Retrieval* 5, no. 970: 203–11.

———. 1975. What are data? *Museum Data Bank Research Report* no. 1.

———. 1977. GRIPHOS. Stony Brook, NY: Museum Computer Network.

———. 1981. Computers and fine arts in the United States. (In) Vezina 1981: 89–98.

———. 1982. Problems in museum networks. (In) *Computers and Cultural Materials Conference* 1982: 119–24.

———. 1984. Identification of objects. (In) Scuola Normale Superiore, vol. 2, 1984: 337–51.

———. 1986. Museum Computer Network in context. (In) Light, Roberts, and Stewart 1986: 37–47.

———. 1987a. Numbers in history. *International Journal of Museum Management and Curatorship* 6, no. 3 (Sep 1987).

———. 1987b. Of quantity and quality. *International Journal of Museum Management and Curatorship* 6, no. 4 (Dec 1987).

Vance, David and Jack Heller. 1971. Structure and content of a museum data bank. *Computers and the Humanities* 6, no. 2 (Nov 1971): 67–84.

Vance, David and IBM Staff. 1973. *Computers in the Museum.* White Plains, NY: International Business Machines Corporation (IBM ref GE20-0406-0).

Varveris, Therese. 1980. *Cataloguer's Manual for the Visual Arts.* Sydney: Australian Art Gallery Directors' Council, Ltd.

Vezina, Raymond, ed. 1981. *Computerized Inventory Standards for Works of Art: Proceedings of a Conference at the Public Archives of Canada, Nov 1–5, 1979.* Montreal: Editions Fides.

Waal, H. van de. 1968. *Decimal Index to the Art of the Low Countries (DAIL).* The Hague: Rijksbureau voor Kunsthistorische Documentatie.

Weiner, Stephen B. 1985. Designing a collections information system for the Smithsonian Institution. *Curator* 28, no. 4 (Dec 1985): 237–48.

Welsh, Peter H. and Steven A. Le Blanc. 1987. Computer literacy and collections management. *Museum News* 65, no. 5 (Jun 1987): 42–51.

Whallon, Robert, Jr. 1971. The computer in archaeology: a critical survey. *Computers and the Humanities* 7, no. 2 (Nov 1971): 29–45.

White, Robert M. 1980. Disk-storage technology. *Scientific American* 243, no. 2 (Aug 1980): 138–48.

Wilhelmi, Lyle. 1983. Microcomputers as systems exhibit controls. *Curator* 26, no. 2 (Jun 1983): 107–20.

Wolters, Christof and Peter-Georg Hausmann. 1982. Objektdokumentation. Berlin: Staatliche Museen Preussischer Kulturbesitz, Institut fuer Museumskunde.

Index

ABOUT THE AUTHORS

ROBERT G. CHENHALL, a former museum director, is presently a consultant to museums and small business in Albuquerque, New Mexico. His previous books include *Museum Cataloging in the Computer Age* and *Nomenclature for Museum Cataloging: A System for Classifying Man-Made Objects*.

DAVID VANCE, as Registrar of The Museum of Modern Art, planned and directed the first computerization of a major art museum catalog. He served for 16 years as President and Executive Director of the Museum Computer Network, Inc.